Easy Math
STEP-BY-STEP

Master High-Frequency Concepts and Skills for Mathematical Proficiency—*FAST!*

Second Edition

William D. Clark, PhD
Sandra Luna McCune, PhD

New York Chicago San Francisco Athens London Madrid Mexico City
Milan New Delhi Singapore Sydney Toronto

1 2 3 4 5 6 7 8 9 LCR 23 22 21 20 19 18

ISBN 978-1-260-13521-3
MHID 1-260-13521-7

e-ISBN 978-1-260-13522-0
e-MHID 1-260-13522-5

Contents

Preface

Easy Math Step-by-Step is an interactive approach to learning basic math. It contains completely worked-out sample solutions that are explained in detailed, step-by-step instructions. Moreover, it features guiding principles, cautions against common errors, and offers other helpful advice as "pop-ups" in the margins. The book takes you from number concepts to skills in elementary algebra and ends with simple descriptive statistics and graphical representation of data. Concepts are broken into basic components to provide ample practice of fundamental skills.

The anxiety you may feel while trying to succeed in math is a real-life phenomenon. Many people experience such a high level of tension when faced with a math problem that they simply cannot perform to the best of their abilities. It is possible to overcome this difficulty by building your confidence in your ability to do math and by minimizing your fear of making mistakes.

No matter how much it might seem to you that math is too hard to master, success will come. Learning math requires lots of practice. Most important, it requires a true confidence in yourself and in the fact that, with practice and persistence, you will be able to say, "I can do this!"

In addition to the many worked-out, step-by-step sample problems, this book presents a variety of exercises and levels of difficulty to provide reinforcement of math concepts and skills. After working a set of exercises, use the worked-out solutions to check your understanding of the concepts.

We sincerely hope *Easy Math Step-by-Step* will help you acquire greater competence and confidence in using math in your future endeavors.

1

Numbers and Operations

In this chapter, you learn about the various sets of numbers that make up the real numbers.

Natural Numbers and Whole Numbers

The *natural numbers* (or *counting numbers*) are the numbers

The three dots indicate that the pattern continues without end.

$$1, 2, 3, 4, 5, 6, 7, 8, \ldots$$

You can represent the natural numbers as equally spaced points on a number line, increasing endlessly in the direction of the arrow, as shown in Figure 1.1.

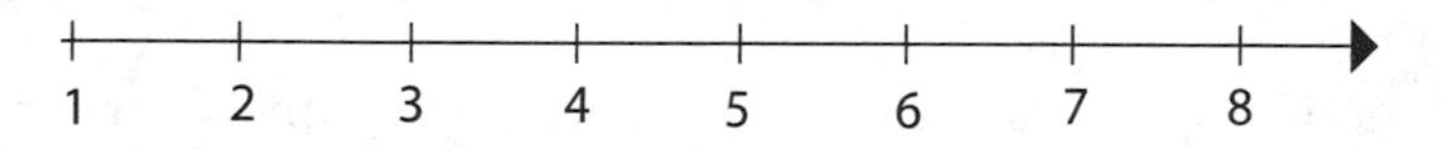

Figure 1.1 Natural numbers

When you include the number 0 with the set of natural numbers, you have the whole numbers:

The number 0 is a whole number, but not a natural number.

$$0, 1, 2, 3, 4, 5, 6, 7, 8, \ldots$$

Like the natural numbers, you can represent the whole numbers as equally spaced points on a number line, increasing endlessly in the direction of the arrow, as shown in Figure 1.2.

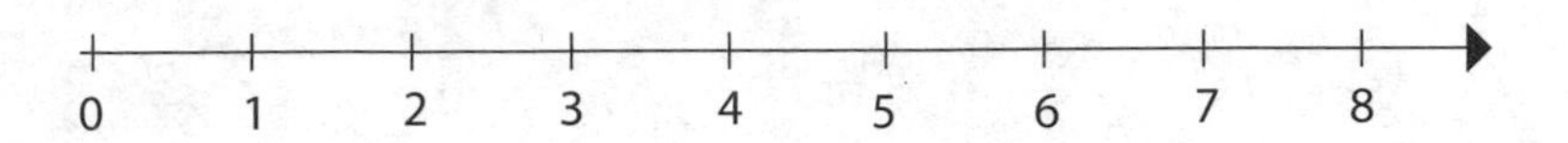

Figure 1.2 Whole numbers

The *graph* of a number is the point on the number line that corresponds to the number, and the number is the *coordinate* of the point. You graph a set of numbers by marking a large dot at each point on the number line corresponding to one of the numbers. The graph of the numbers 2, 3, and 7 is shown in Figure 1.3.

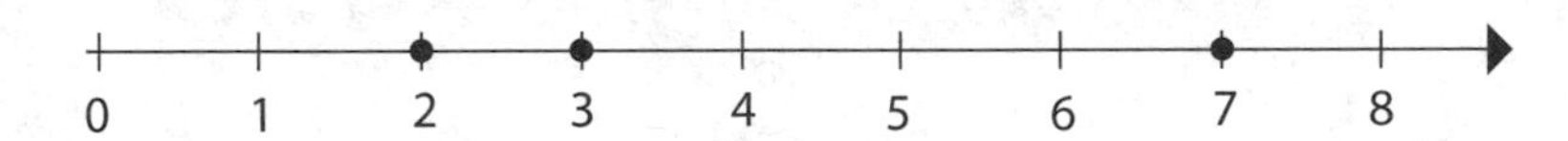

Figure 1.3 Graph of 2, 3, and 7

Integers

On the number line shown in Figure 1.4, the point 1 unit to the left of 0 corresponds to the number –1 (read as "negative one"), the point 2 units to the left of 0 corresponds to the number –2, the point 3 units to the left of 0 corresponds to the number –3, and so on. The number –1 is the opposite of 1, –2 is the opposite of 2, –3 is the opposite of 3, and so on. The number 0 is its own opposite.

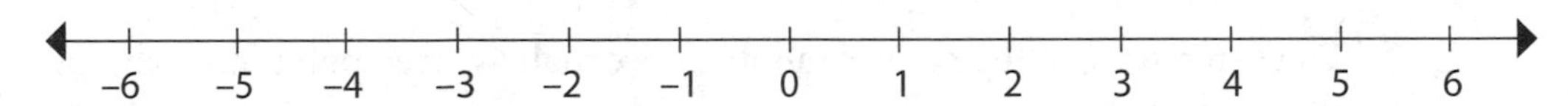

Figure 1.4 Whole numbers and their opposites

A number and its *opposite* are exactly the same distance from 0. For instance, 3 and –3 are opposites, and each is 3 units from 0.

The whole numbers and their opposites make up the *integers*:

> The number 0 is neither positive nor negative.

$$\ldots, -3, -2, -1, 0, 1, 2, 3, \ldots$$

> It is not necessary to write a + sign on positive numbers (although it's not wrong to do so). If no sign is written, then you know the number is positive.

The integers are either *positive* (1, 2, 3, ...), *negative* (..., –3, –2, –1), or 0. Positive numbers are located to the right of 0 on the number line, and negative numbers are to the left of 0, as shown in Figure 1.5.

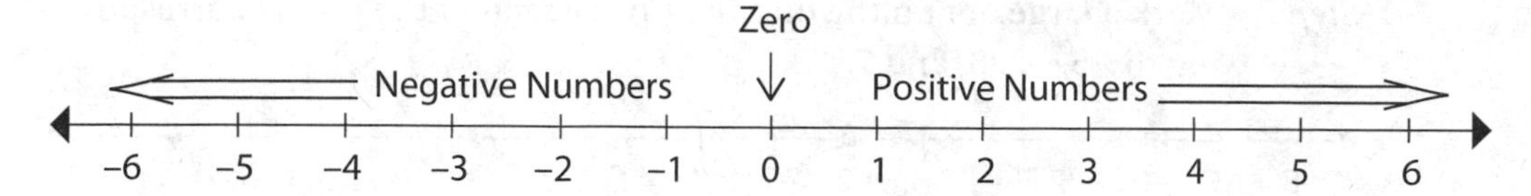

Figure 1.5 Number line

Problem Find the opposite of the given number.

a. 8

b. −4

> A nonzero number's opposite can be positive or negative. The opposite of a positive number is negative. The opposite of a negative number is positive.

Solution

a. 8

Step 1. Describe the location of 8 and its opposite on a number line.

8 is 8 units to the right of 0. The opposite of 8 is 8 units to the left of 0.

Step 2. Specify the opposite of 8.

The number that is 8 units to the left of 0 is −8. Therefore, −8 is the opposite of 8.

b. −4

Step 1. Describe the location of −4 and its opposite on a number line.

−4 is 4 units to the left of 0. The opposite of −4 is 4 units to the right of 0.

Step 2. Specify the opposite of −4.

The number that is 4 units to the right of 0 is 4. Therefore, 4 is the opposite of −4.

Problem Graph the integers −5, −2, 3, and 7.

Solution

Step 1. Draw a number line.

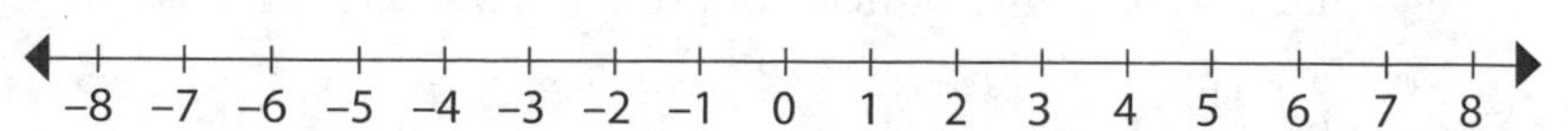

Step 2. Mark a large dot on the number line at each of the points corresponding to –5, –2, 3, and 7.

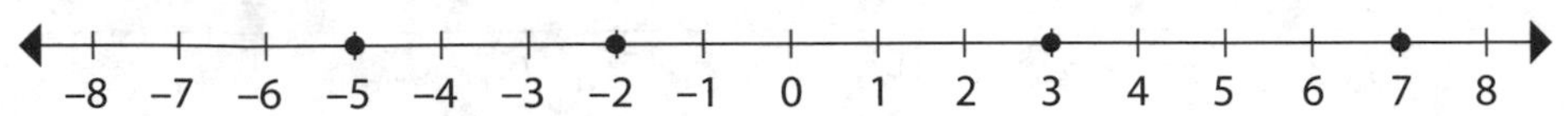

Rational, Irrational, and Real Numbers

The number $\frac{1}{4}$ is an example of a rational number. A *rational number* is a number that can be expressed as a quotient of an integer divided by an integer other than 0. That is, the rational numbers are all the numbers that can be expressed as

> The number 0 is excluded as a denominator for $\frac{p}{q}$ because division by 0 is undefined, so $\frac{p}{0}$ has no meaning no matter what number you put in the place of *p*.

$$\frac{p}{q}, \text{ where } p \text{ and } q \text{ are integers, } q \neq 0$$

The rational numbers include positive and negative fractions, decimals, and percents. All of the natural numbers, whole numbers, and integers are rational numbers as well because you can express each of these numbers, as shown here.

> The quotient of two integers, where the divisor is not zero, is a *ratio*. Hence, every rational number is the ratio of two integers.

$$\ldots, -3 = \frac{-3}{1}, -2 = \frac{-2}{1}, -1 = \frac{-1}{1}, 0 = \frac{0}{1}, 1 = \frac{1}{1}, 2 = \frac{2}{1}, 3 = \frac{3}{1}, \ldots$$

The decimal representations of rational numbers terminate or repeat. For instance, $\frac{1}{4} = 0.25$ is a rational number whose decimal representation terminates, and $\frac{2}{3} = 0.666\ldots$ is a rational number whose decimal representation repeats. You can show a repeating decimal by placing a line over the block of digits that repeats, like this: $\frac{2}{3} = 0.\overline{6}$. You also might find it convenient to round the repeating decimal to a certain number of decimal places. For instance, rounded to two decimal places, $\frac{2}{3} \approx 0.67$.

> The symbol ≈ means "is approximately equal to."

Note: Fractions, decimals, and percents are discussed at length in Chapters 5–7.

The *irrational numbers* are numbers whose decimal representations neither terminate nor repeat. These numbers cannot be expressed as ratios of two integers. For instance, the positive number that multiplies by itself to give 2 is an irrational number called the positive square root of 2. You use the square root radical symbol ($\sqrt{\ }$) to show the positive square root of 2 like this: $\sqrt{2}$. Every positive number has two square roots: a positive square root and a negative square root. The other square root of 2 is $-\sqrt{2}$. It also is an irrational number. The number 0 has only one square root, namely 0 (which is a rational number). (See Chapter 10 for an additional discussion of square roots.)

You cannot express $\sqrt{2}$ as the ratio of two integers, nor can you express it precisely in decimal form. Its decimal equivalent continues on and on without a pattern of any kind, so no matter how far you go with decimal places, you can only approximate $\sqrt{2}$. For instance, rounded to three decimal places, $\sqrt{2} \approx 1.414$. Do not be misled, however, because even though you cannot determine an exact value for $\sqrt{2}$, it is a number that occurs frequently in the real world. For instance, designers and builders encounter $\sqrt{2}$ as the length of the diagonal of a square that has sides with length of 1 unit, as shown in Figure 1.6.

Not all roots are irrational. For instance, $\sqrt{36} = 6$.

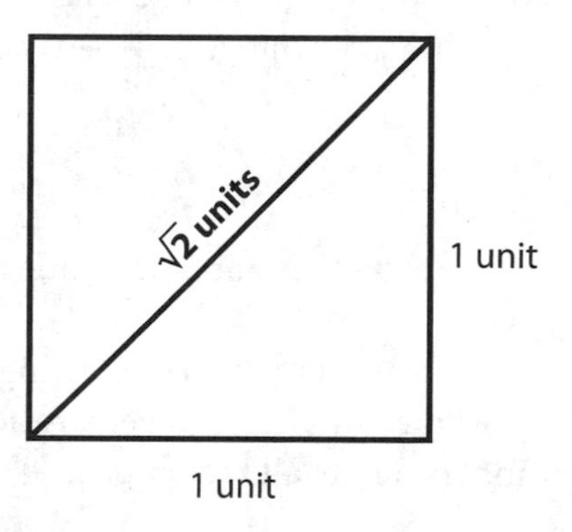

Figure 1.6 Diagonal of a unit square

There are infinitely many square roots and other roots as well that are irrational.

Two important irrational numbers are the numbers represented by the

Although, in the past, you might have used 3.14 or $\frac{22}{7}$ for π, π does not equal either of these numbers. The numbers 3.14 and $\frac{22}{7}$ are rational numbers, but π is irrational.

symbols π and e. The number π is the ratio of the circumference of a circle to its diameter, and the number e is used extensively in calculus. Most scientific and graphing calculators have π and e keys. To nine-decimal-place accuracy, $\pi \approx 3.141592654$ and $e \approx 2.718281828$.

The *real numbers* are all the rational and irrational numbers put together. They are all the numbers on the number line (see Figure 1.7). Every point on the number line corresponds to a real number, and every real number corresponds to a point on the number line.

Be careful: Square roots of negative numbers are not real numbers. This fact is true because no real number times itself yields a negative product.

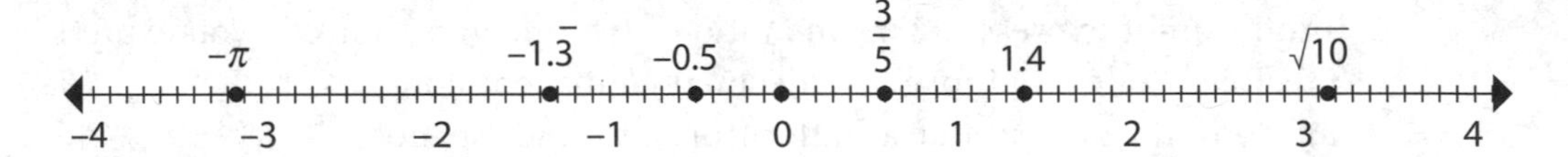

Figure 1.7 Real number line

The relationship among the various sets of numbers included in the real numbers is shown in Figure 1.8.

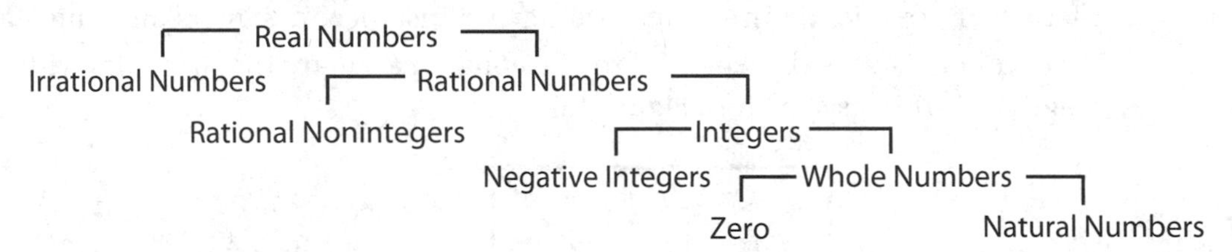

Figure 1.8 Real numbers

Note: Hereafter in this book, all numbers are understood to be real numbers.

Problem Categorize the given number according to its membership in the natural numbers, whole numbers, integers, rational numbers, irrational numbers, or real numbers. (State all that apply.)

a. 0

b. 0.75

c. −25

d. $\sqrt{36}$

e. −0.35

f. $\frac{2}{3}$

g. $-\frac{162}{341}$

h. $\sqrt{17}$

Solution

Step 1. Recall the characteristics of the natural numbers, whole numbers, integers, rational numbers, irrational numbers, and real numbers.

Step 2. Categorize the number.

a. 0 is a whole number, an integer, a rational number, and a real number.

b. 0.75 is a rational number and a real number.

c. −25 is an integer, a rational number, and a real number.

d. $\sqrt{36}$ is equal to 6, which is a natural number, a whole number, an integer, a rational number, and a real number.

e. −0.35 is a rational number and a real number.

f. $\frac{2}{3}$ is a rational number and a real number.

g. $-\frac{162}{341}$ is a rational number and a real number.

h. $\sqrt{17}$ is an irrational number and a real number.

Problem Graph the numbers $-4, -2.5, 0, \frac{3}{4}$, and 3.6.

Solution

Step 1. Draw a number line.

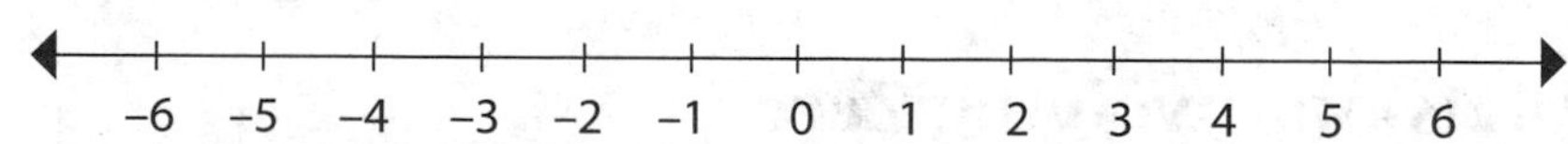

Step 2. Mark a large dot on the number line at each of the points corresponding to $-4, -2.5, 0, \frac{3}{4}$, and 3.6.

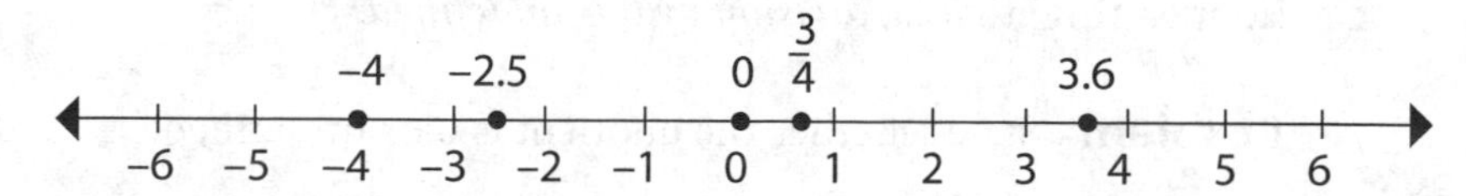

Terminology for the Four Basic Operations

For the math you do in your everyday world, you work with the real numbers. Addition, subtraction, multiplication, and division are the four basic operations you use. Each of the operations has special symbolism and terminology associated with it. Table 1.1 shows the terminology and symbolism for the operations.

Table 1.1 The Four Basic Operations

OPERATION	SYMBOLS(S) USED	NAME OF PARTS	EXAMPLE
Addition	+ (plus sign)	addend + addend = sum	$6+9=15$
Subtraction	– (minus sign)	sum – addend = difference	$15-9=6$
Multiplication	× (times sign)	factor × factor = product	$10\times 2=20$
	· (raised dot)	factor · factor = product	$10\cdot 2=20$
	( )( ) parentheses	(factor)(factor) = product	$(10)(2)=20$
Division	÷ (division sign)	dividend ÷ divisor = quotient	$20\div 2=10$
	$\overline{)\ }$ (long division symbol)	$\text{divisor}\overset{\text{quotient}}{\overline{)\text{dividend}}}$	$10\overset{2}{\overline{)20}}$
	fraction bar	$\dfrac{\text{dividend}}{\text{divisor}}=\text{quotient}$	$\dfrac{20}{10}=2$

As you can see from the examples in Table 1.1, addition and subtraction "undo" each other. Similarly, multiplication and division undo each other, *as long as no division by 0 occurs.*

Note: A number next to a parentheses means that the quantity inside the parentheses is to be multiplied by the number. For example, 3(10 + 5) means 3 times the sum of 10 and 5.

Division Involving Zero

You must be *very* careful when you have zero in a division problem. The number 0 can be the dividend, provided the divisor is not 0—the quotient will be 0. But 0 can *never* be the divisor. The quotient of any number divided by 0 has no meaning; that is, *division by 0 is undefined.*

Problem State whether the quotient is 0 or undefined.

a. $9\div 0$

b. $0\div 9$

c. $\frac{0}{-5}$

d. $\frac{-5}{0}$

e. $\frac{0}{0}$

Solution

a. $9 \div 0$

Step 1. Recall that division by 0 is undefined, so state that the quotient is undefined.

$9 \div 0$ is undefined.

b. $0 \div 9$

Step 1. Recall that the quotient is 0 when 0 is divided by a nonzero number, so state that the quotient is 0.

$0 \div 9 = 0$

c. $\frac{0}{-5}$

Step 1. Recall that the quotient is 0 when 0 is divided by a nonzero number, so state that the quotient is 0.

$\frac{0}{-5} = 0$

d. $\frac{-5}{0}$

Step 1. Recall that division by 0 is undefined, so state that the quotient is undefined.

$\frac{-5}{0}$ is undefined.

e. $\frac{0}{0}$

Step 1. Recall that division by 0 is undefined, so state that the quotient is undefined.

$\frac{0}{0}$ is undefined.

Properties of Real Numbers

The real numbers have the following 11 properties under the operations of addition and multiplication.

For all real numbers, you have:

1. **Closure Property of Addition.** This property guarantees that the sum of any two real numbers is always a real number.

   Examples

   $(4+5)$ is a real number.

   $\left(\frac{1}{2}+\frac{3}{4}\right)$ is a real number.

   $(0.54+6.1)$ is a real number.

2. **Closure Property of Multiplication.** This property guarantees that the product of any two real numbers is always a real number.

   Examples

   (2×7) is a real number.

   $\left(\frac{1}{3}\times\frac{5}{8}\right)$ is a real number.

   $(2.5)(10.35)$ is a real number.

3. **Commutative Property of Addition.** This property allows you to reverse the order of the numbers when you add, without changing the sum.

   Examples

   $4+5=5+4=9$

$$\frac{1}{2}+\frac{3}{4}=\frac{3}{4}+\frac{1}{2}=\frac{5}{4}$$

$$0.54+6.1=6.1+0.54=6.64$$

4. **Commutative Property of Multiplication.** This property allows you to reverse the order of the numbers when you multiply, without changing the product.

   Examples

   $$2\times7=7\times2=14$$

   $$\frac{1}{3}\times\frac{5}{8}=\frac{5}{8}\times\frac{1}{3}=\frac{5}{24}$$

   $$(2.5)(10.35)=(10.35)(2.5)=25.875$$

5. **Associative Property of Addition.** This property says that when you have three numbers to add together, the final sum will be the same regardless of the way you group the numbers (two at a time in the same order) to perform the addition.

   Example

   Suppose you want to compute $6+3+7$. In the order given, you have two ways to group the numbers for addition:

   $$(6+3)+7=9+7=16 \text{ or } 6+(3+7)=6+10=16$$

   Either way, 16 is the final sum.

6. **Associative Property of Multiplication.** This property says that when you have three numbers to multiply together, the final product will be the same regardless of the way you group the numbers (two at a time in the same order) to perform the multiplication.

   Example

   Suppose you want to compute $7\times2\times\frac{1}{2}$. In the order given, you have two ways to group the numbers for multiplication:

   $$(7\times2)\times\frac{1}{2}=14\times\frac{1}{2}=7 \text{ or}$$

   $$7\times\left(2\times\frac{1}{2}\right)=7\times1=7$$

   Either way, 7 is the final product.

> The associative property is needed when you have to add or multiply more than two numbers, because you can do addition or multiplication on only two numbers at a time. Thus, when you have three numbers, you must decide which two numbers you want to start with—the first two or the last two (assuming you keep the same order). Either way, your final answer is the same.

7. **Additive Identity Property.** This property guarantees that you have the number 0 (called the *additive identity*) for which its sum with any number is the number itself.

Examples

$$8+0=8 \text{ and } 0+8=8$$

$$\frac{5}{6}+0=\frac{5}{6} \text{ and } 0+\frac{5}{6}=\frac{5}{6}$$

8. **Multiplicative Identity Property.** This property guarantees that you have the number 1 (called the *multiplicative identity*) for which its product with any number is the number itself.

Examples

$$5\times 1=5 \text{ and } 1\times 5=5$$

$$\frac{7}{8}\times 1=\frac{7}{8} \text{ and } 1\times\frac{7}{8}=\frac{7}{8}$$

9. **Additive Inverse Property.** This property guarantees that every number has an opposite (called its *additive inverse*) whose sum with the number is 0.

Examples

$$6+-6=0 \text{ and } -6+6=0$$

$$7.43+-7.43=0 \text{ and } -7.43+7.43=0$$

10. **Multiplicative Inverse Property.** This property guarantees that every number, *except 0*, has a reciprocal (called its *multiplicative inverse*) whose product with the number is 1.

Example

$$3\times\frac{1}{3}=\frac{1}{3}\times 3=1$$

11. **Distributive Property.** This property says that when you have a number times a sum, you can either add first and then multiply or multiply first and then add. Either way, the final answer is the same.

Example

$3(10+5)$ can be computed two ways:

add first to obtain $3(10 + 5) = 3 \cdot 15 = 45$ or multiply first to obtain $3(10 + 5) = 3 \cdot 10 + 3 \cdot 5 = 30 + 15 = 45$

Either way, the answer is 45.

The distributive property is the only property that involves both addition and multiplication at the same time. Another way to express the distributive property is to say that *multiplication distributes over addition.*

Subtraction and division are not mentioned in the properties listed, because you can always turn subtraction into addition by "adding the opposite," and you can turn division by a nonzero number into multiplication by "multiplying by the reciprocal." (See "Subtracting Signed Numbers" and "Dividing Signed Numbers" in Chapter 2 for further clarification.)

When you subtract a number, you get the same answer as you do when you add its opposite.

When you divide by a nonzero number, you get the same answer as you do when you multiply by its reciprocal.

Problem Identify the property illustrated.

a. $0+1.25=1.25$

b. $(3+4.5)$ is a real number

c. $\frac{3}{4}\cdot\frac{5}{6}=\frac{5}{6}\cdot\frac{3}{4}$

Solution

Step 1. Recall the 11 properties: closure property of addition, closure property of multiplication, commutative property of addition, commutative property of multiplication, associative property of addition, associative property of multiplication, additive identity property, multiplicative identity property, additive inverse property, multiplicative inverse property, and distributive property.

a. $0+1.25=1.25$

Step 2. Identify the property illustrated.

additive identity property

b. $(3+4.5)$ is a real number

Step 2. Identify the property illustrated.

closure property of addition

c. $\frac{3}{4}\cdot\frac{5}{6}=\frac{5}{6}\cdot\frac{3}{4}$

Step 2. Identify the property illustrated.

commutative property of multiplication

Besides the 11 properties given, the number 0 has the following unique characteristic.

Zero Factor Property

If a number is multiplied by 0, then the product is 0; if the product of two numbers is 0, then at least one of the numbers is 0.

Problem Find the product.

a. $-9\cdot 0$

b. $0\cdot\frac{15}{100}$

Solution

a. $-9\cdot 0$

Step 1. Given that 0 is a factor of the product, apply the zero factor property.

$$-9\cdot 0=0$$

b. $0\cdot\frac{15}{100}$

Step 1. Given that 0 is a factor of the product, apply the zero factor property.

$$0\cdot\frac{15}{100}=0$$

This property explains why 0 does not have a reciprocal. There is no number that multiplies by 0 to give 1—because any number multiplied by 0 is 0.

Exercise 1

For 1–9, list all the sets of the real number system to which the given number belongs. (State all that apply.)

1. 10
2. -7.3
3. -74
4. $-1{,}000$
5. 0.555...
6. $-\frac{3}{4}$
7. $\frac{345}{63}$
8. 0
9. $\sqrt{15}$

For 10–12, state whether the quotient is 0 or undefined.

10. $0 \div 0$
11. $\frac{0}{50}$
12. $\frac{65}{0}$

For 13–15, identify the property illustrated.

13. $4(10 + 3) = 4 \cdot 10 + 4 \cdot 3$
14. $35 \times 0 = 0$
15. $\frac{3}{5} \times \left(\frac{5}{6} \times 6\right) = \left(\frac{3}{5} \times \frac{5}{6}\right) \times 6$

2

Integers

In this chapter, you learn about performing operations with integers. Before proceeding with addition, subtraction, multiplication, and division of integers, the discussion begins with comparing and finding the absolute value of numbers.

Comparing Integers

Recall that the integers are the numbers

$$\ldots, -3, -2, -1, 0, 1, 2, 3, \ldots$$

Comparing numbers uses the *inequality symbols* shown in Table 2.1.

Table 2.1 Inequality Symbols

INEQUALITY SYMBOL	*EXAMPLE*	*READ AS*
$<$	$2 < 7$	"2 is less than 7"
$>$	$7 > 2$	"7 is greater than 2"
$\leq$	$9 \leq 9$	"9 is less than or equal to 9"
$\geq$	$10 \geq 4$	"10 is greater than or equal to 4"
$\neq$	$2 \neq 7$	"2 is not equal to 7"

The statement "9 is less than or equal to 9" is true because 9 equals 9.

Graphing the numbers on a number line is helpful when you compare two numbers. If the numbers coincide, then they are equal. Otherwise, they are unequal, and the number that is farther to the right is the greater number.

Problem Which is greater –7 or –2?

Solution

Step 1. Graph –7 and –2 on a number line.

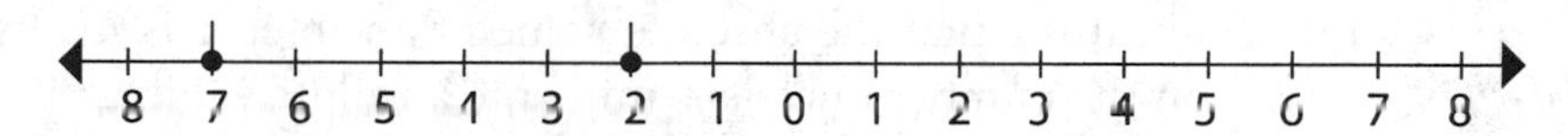

Step 2. Identify the number that is farther to the right as the greater number.

–2 is to the right of –7, so –2 > –7.

Absolute Value

The concept of absolute value plays an important role in computations with signed numbers. The *absolute value* of an integer is its distance from 0 on the number line. For example, as shown in Figure 2.1, the absolute value of –8 is 8 because –8 is 8 units from 0.

Absolute value is a distance, so it is *never* negative.

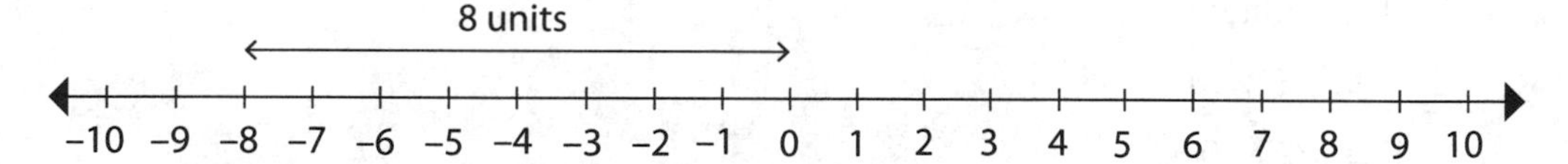

Figure 2.1 Absolute value of –8

You indicate the absolute value of a number by placing the number between a pair of vertical bars like this: $|-8|$ (read as "the absolute value of negative eight"). Thus, $|-8| = 8$.

Problem Find the indicated absolute value.

a. $|-30|$

b. $|4|$

c. $|-2|$

Solution

a. $|-30|$

Step 1. Recalling that the absolute value of an integer is its distance from 0 on the number line, determine the absolute value.

$|-30| = 30$ because -30 is 30 units from 0 on the number line.

b. $|4|$

Step 1. Recalling that the absolute value of an integer is its distance from 0 on the number line, determine the absolute value.

$|4| = 4$ because 4 is 4 units from 0 on the number line.

c. $|-2|$

Step 1. Recalling that the absolute value of an integer is its distance from 0 on the number line, determine the absolute value.

$|-2| = 2$ because -2 is 2 units from 0 on the number line.

> As you likely noticed, the absolute value of a number is the value of the number with no sign attached. This strategy works for a number whose value you know (that is, a number that you can locate on a number line), but do not use it when you don't know the value of the number.

Problem Which number has the greater absolute value?

a. –35, 60

b. 35,–60

c. –7, 2

d. –21, 17

Solution

a. –35, 60

Step 1. Determine the absolute values.

$|-35| = 35$, $|60| = 60$

Step 2. Compare the absolute values.

60 has the greater absolute value because $60 > 35$.

b. 35,–60

Step 1. Determine the absolute values.

$|35| = 35, \; |-60| = 60$

Step 2. Compare the absolute values.

–60 has the greater absolute value because $60 > 35$.

c. –7, 2

Step 1. Determine the absolute values.

$|-7| = 7, \; |2| = 2$

Step 2. Compare the absolute values.

–7 has the greater absolute value because $7 > 2$.

d. –21, 17

Step 1. Determine the absolute values.

$|-21| = 21, \; |17| = 17$

Step 2. Compare the absolute values.

–21 has the greater absolute value because $21 > 17$.

Don't make the mistake of trying to compare the numbers without first finding the absolute values.

Integers are called signed numbers because these numbers may be positive, negative, or 0. From your knowledge of arithmetic, you already know how to do addition, subtraction, multiplication, and division with positive numbers and 0. To do these operations with all signed numbers, you simply use the absolute values of the numbers and follow these eight rules.

Adding Signed Numbers

Addition of Signed Numbers

Rule 1. To add two numbers that have the same sign, add their absolute values and give the sum their common sign.

Rule 2. To add two numbers that have opposite signs, subtract the lesser absolute value from the greater absolute value and give the sum the sign of the number with the greater absolute value; if the two numbers have the same absolute value, their sum is 0.

These rules might sound complicated, but practice will make them your own. One helpful hint is that when you need the absolute value of a specific number, just use the value of the number with no sign attached.

Rule 3. The sum of 0 and any number is the number.

The number line is a good tool for illustrating the addition of signed numbers, as shown below.

Problem Add the two numbers. Illustrate the addition on a number line.

a. 4 and −7

b. −5 and 7

c. −3 and −4

Solution

a. 4 and −7

Step 1. Begin at 0 and move 4 units in the positive direction to 4.

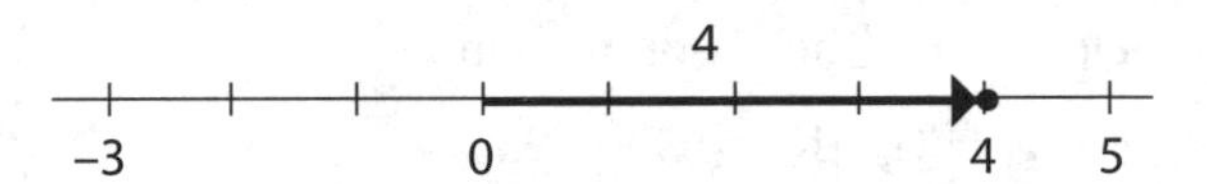

Step 2. From that point, move 7 units in the negative direction to −3.

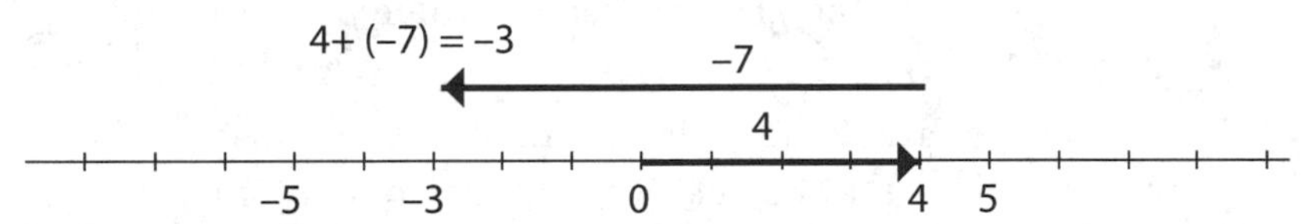

Step 3. Express the result.

$4 + (-7) = -3$

If you find it helpful, you can put parentheses around negative numbers for clarity.

b. −5 and 7

Step 1. Start at 0 and go 5 units in the negative direction to −5.

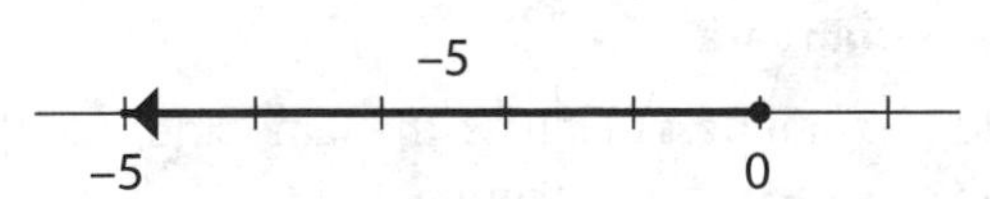

Step 2. From that point, go 7 units in the positive direction to 2.

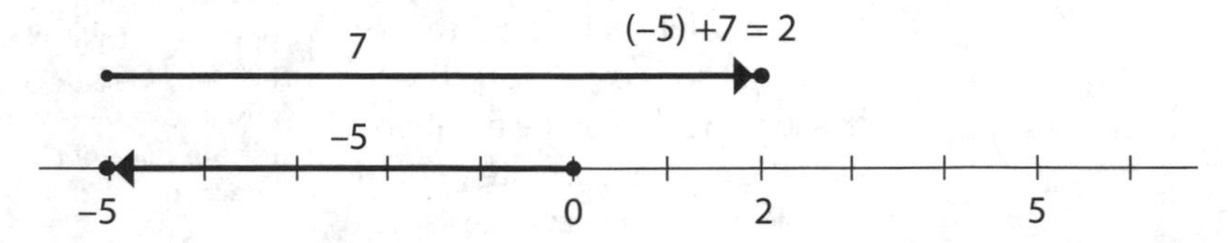

Step 3. Express the result.

$-5+7=2$

c. -3 and -4

Step 1. Start at 0 and move 3 units in the negative direction to -3.

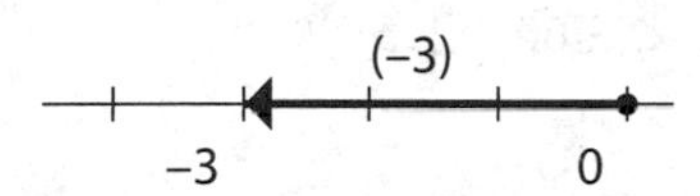

Step 2. From that point, move 4 units in the negative direction to -7.

(−3) + (−4) = −7 (−4) (−3)

−5 −3 0

Step 3. Express the result.

$(-3)+(-4)=-7$

Problem Find the sum.

a. $-35+-60$

b. $35+-60$

c. $-35+60$

Draw number line illustrations for parts d and e.

d. $3+(-3)$

e. $-5+2$

Solution

a. $-35+-60$

Step 1. Determine which addition rule applies.

$-35+-60$

The signs are the same (both negative), so use Rule 1.

Step 2. Add the absolute values, 35 and 60.

$35+60=95$

Step 3. Give the sum a negative sign (the common sign).

$-35+-60=-95$

b. $35 + -60$

Step 1. Determine which addition rule applies.

$35 + -60$

The signs are opposites (one positive and one negative), so use Rule 2.

Step 2. Subtract 35 from 60 because $|-60| > |35|$.

$60 - 35 = 25$

Step 3. Make the sum negative because -60 has the greater absolute value.

$35 + -60 = -25$

c. $-35 + 60$

Step 1. Determine which addition rule applies.

$-35 + 60$

The signs are opposites (one negative and one positive), so use Rule 2.

Step 2. Subtract 35 from 60 because $|60| > |-35|$.

$60 - 35 = 25$

Step 3. Keep the sum positive because 60 has the greater absolute value.

$-35 + 60 = 25$

d. $3 + (-3)$

Step 1. Start at 0 and go 3 units in the positive direction to 3.

0 3

Step 2. From that point, go 3 units in the negative direction to 0.

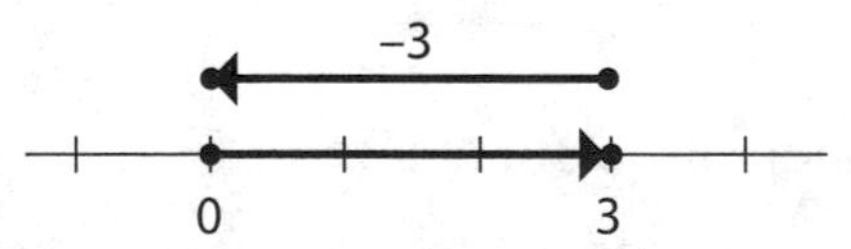

Step 3. Express the result.

$3 + (-3) = 0$

e. $-5+2$

Step 1. Start at 0 and go 5 units in the negative direction to −5.

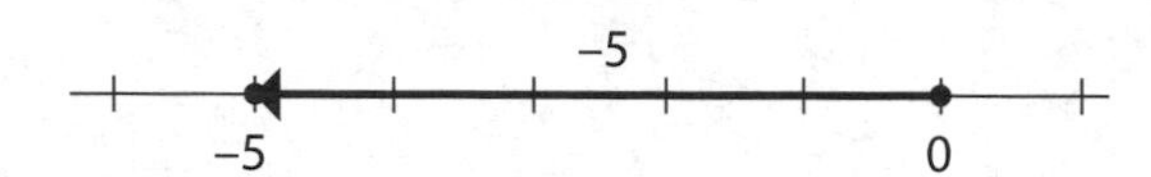

Step 2. From that point, go 2 units in the positive direction to −3.

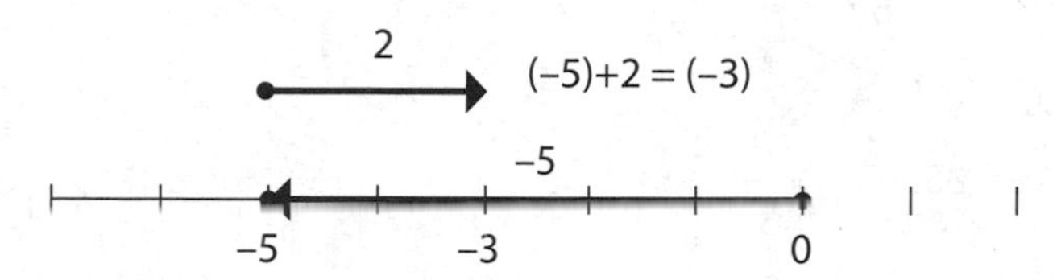

Step 3. Express the result.

$-5+2=-3$

Subtracting Signed Numbers

You subtract signed numbers by changing the subtraction problem in a special way to an addition problem, so that you can apply the rules for addition of signed numbers. Here is the rule.

Subtraction of Signed Numbers

Rule 4. To subtract two numbers, keep the first number and add the opposite of the second number.

To apply this rule, think of the minus sign, −, as "add the opposite of." In other words, "subtracting a number" and "adding the opposite of the number" give the same answer.

Problem Change the subtraction problem to an addition problem.

a. $-35-60$

b. $35-60$

c. $60-35$

d. $-35-(-60)$

e. $0-60$

f. $-60-0$

Solution

a. $-35-60$

Step 1. Keep -35.

-35

Step 2. Add the opposite of 60.

$=-35+-60$

b. $35-60$

Step 1. Keep 35.

35

Step 2. Add the opposite of 60.

$=35+-60$

c. $60-35$

Step 1. Keep 60.

60

Step 2. Add the opposite of 35.

$=60+-35$

d. $-35-(-60)$

Step 1. Keep -35.

-35

Step 2. Add the opposite of -60.

$=-35+60$

e. $0-60$

Step 1. Keep 0.

0

Step 2. Add the opposite of 60.

$=0+-60$

f. $-60 - 0$

Step 1. Keep -60.

-60

Step 2. Add the opposite of 0.

$= -60 + 0$

Remember that 0 is its own opposite.

Problem Compute as indicated.

a. $-35 - 60$

b. $35 - 60$

c. $60 - 35$

d. $-35 - (-60)$

e. $0 - 60$

f. $-60 - 0$

Solution

a. $-35 - 60$

Step 1. Keep -35 and add the opposite of 60.

$-35 - 60$

$= -35 + -60$

Step 2. The signs are the same (both negative), so use Rule 1 for addition.

$= -95$

Step 3. Review the main results.

$-35 - 60$

$= -35 + -60$

$= -95$

Reviewing your work is a smart habit to cultivate. You will catch many careless errors with this strategy.

b. $35 - 60$

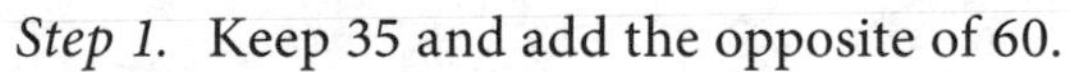

Step 1. Keep 35 and add the opposite of 60.

$35 - 60$

$= 35 + -60$

Step 2. The signs are opposites (one positive and one negative), so use Rule 2 for addition.

$= -25$

Step 3. Review the main results.

$35 - 60$

$= 35 + -60$

$= -25$

c. $60 - 35$

Step 1. Keep 60 and add the opposite of 35.

$60 - 35$

$= 60 + -35$

Step 2. The signs are opposites (one positive and one negative), so use Rule 2 for addition.

$= 25$

Step 3. Review the main results.

$60 - 35$

$= 60 + -35$

$= 25$

d. $-35 - (-60)$

Step 1. Keep -35 and add the opposite of -60.

$-35 - (-60)$

$= -35 + 60$

Step 2. The signs are opposites (one positive and one negative), so use Rule 2 for addition.

$= 25$

Step 3. Review the main results.

$-35 - (-60)$

$-35 + 60$

$= 25$

e. $0 - 60$

Step 1. Keep 0 and add the opposite of 60.

$$0 - 60$$
$$= 0 + -60$$

Step 2. 0 is added to a number, so the sum is the number (Rule 3 for addition).

$$= -60$$

Step 3. Review the main results.

$$0 - 60$$
$$= 0 + -60$$
$$= -60$$

f. $-60 - 0$

Step 1. Keep -60 and add the opposite of 0.

$$-60 - 0$$
$$= -60 + 0$$

Step 2. 0 is added to a number, so the sum is the number (Rule 3 for addition).

$$= -60$$

Step 3. Review the main results.

$$-60 - 0$$
$$= -60 + 0$$
$$= -60$$

Notice that subtraction is *not* commutative. That is, in general, for real numbers a and b, $a - b \neq b - a$.

Use of the – Symbol

Before going on, it is important that you distinguish the various uses of the short horizontal – symbol. Thus far, this symbol has three uses: (1) as part of a number to show that the number is negative, (2) as an indicator to find the opposite of the number that follows, and (3) as the minus sign indicating subtraction.

Problem Given the statement $\underset{(1)}{\underset{\uparrow}{-}}(\underset{(2)}{\underset{\uparrow}{-}}35)\underset{(3)}{\underset{\uparrow}{-}}60 = 35 + \underset{(4)}{\underset{\uparrow}{-}}60$

a. Describe the use of the – symbols at (1), (2), (3), and (4).

b. Express the statement $-(-35)-60=35+-60$ in words.

Solution

a. Describe the use of the – symbols at (1), (2), (3), and (4).

Step 1. Interpret each – symbol.

The – symbol at (1) is an indicator to find the opposite of –35.

The – symbol at (2) is part of the number –35 that shows –35 is negative.

The – sign at (3) is the minus sign indicating subtraction.

The – symbol at (4) is part of the number –60 that shows –60 is negative.

Don't make the error of referring to negative numbers as "minus numbers."

The minus sign always has a number to its immediate left.

There is never a number to the immediate left of a negative sign.

b. Express the statement $-(-35)-60=35+-60$ in words.

Step 1. Translate the statement into words.

$-(-35)-60=35+-60$ is read as "the opposite of negative thirty-five minus sixty is thirty-five plus negative sixty."

Multiplying Signed Numbers

For multiplication of signed numbers, use the following three rules.

Multiplication of Signed Numbers

Rule 5. To multiply two numbers that have the same sign, multiply their absolute values and keep the product positive.

Rule 6. To multiply two numbers that have opposite signs, multiply their absolute values and make the product negative.

Rule 7. The product of 0 and any number is 0.

Unlike in addition, when you multiply two positive or two negative numbers, the product is positive no matter what. Similarly, unlike in addition, when you multiply two numbers that have opposite signs, the product is negative—it doesn't matter which number has the greater absolute value.

Problem Find the product.

a. $(-3)(-40)$

b. $(3)(40)$

c. $(-3)(40)$

d. $(3)(-40)$

e. $(358)(0)$

Solution

a. $(-3)(-40)$

Step 1. Determine which multiplication rule applies.

$(-3)(-40)$

The signs are the same (both negative), so use Rule 5.

Step 2. Multiply the absolute values, 3 and 40.

$(3)(40) = 120$

Step 3. Keep the product positive.

$(-3)(-40) = 120$

b. $(3)(40)$

Step 1. Determine which multiplication rule applies.

$(3)(40)$

The signs are the same (both positive), so use Rule 5.

Notice that no real number times itself is a negative number. According to Rule 5, if a real number is positive, the product of that number times itself is positive; and, if a real number is negative, the product of that number times itself is positive as well. For example, $(6)(6) = 36$ and $(-6)(-6) = 36$.

Step 2. Multiply the absolute values, 3 and 40.

$(3)(40) = 120$

Step 3. Keep the product positive.

$(3)(40) = 120$

c. $(-3)(40)$

Step 1. Determine which multiplication rule applies.

$(-3)(40)$

The signs are opposites (one negative and one positive), so use Rule 6.

Step 2. Multiply the absolute values, 3 and 40.

$$(3)(40) = 120$$

Step 3. Make the product negative.

$$(-3)(40) = -120$$

d. (3)(–40)

Step 1. Determine which multiplication rule applies.

$$(3)(-40)$$

The signs are opposites (one positive and one negative), so use Rule 6.

Step 2. Multiply the absolute values, 3 and 40.

$$(3)(40) = 120$$

Step 3. Make the product negative.

$$(3)(-40) = -120$$

e. (358)(0)

Step 1. Determine which multiplication rule applies.

$$(358)(0)$$

0 is one of the factors, so use Rule 7.

Step 2. Find the product.

$$(358)(0) = 0$$

Rules 5 and 6 tell you how to multiply two nonzero numbers, but often you will want to find the product of more than two numbers. To do this, multiply in pairs. You can keep track of the sign as you go along, or you simply can use the following guideline:

When 0 is one of the factors, the product is *always* 0; otherwise, products that have an even number of *negative* factors are positive, whereas, those that have an odd number of *negative* factors are negative.

Notice that if there is no zero factor, then the sign of the product is determined by how many *negative* factors you have.

Problem Find the product.

a. $(600)(-40)(-1{,}000)(0)(-30)$

b. $(-3)(-10)(-5)(25)(-1)(-2)$

c. $(-2)(-4)(-10)(1)(-20)$

Solution

a. $(600)(-40)(-1{,}000)(0)(-30)$

Step 1. 0 is one of the factors, so the product is 0.

$$(600)(-40)(-1{,}000)(0)(-30) = 0$$

b. $(-3)(-10)(-5)(25)(-1)(-2)$

Step 1. Find the product ignoring the signs.

$$(3)(10)(5)(25)(1)(2) = 7{,}500$$

Step 2. You have five negative factors, so make the product negative.

$$(-3)(-10)(-5)(25)(-1)(-2) = -7{,}500$$

c. $(-2)(-4)(-10)(1)(-20)$

Step 1. Find the product ignoring the signs.

$$(2)(4)(10)(1)(20) = 1{,}600$$

Step 2. You have four negative factors, so leave the product positive.

$$(-2)(-4)(-10)(1)(-20) = 1{,}600$$

Dividing Signed Numbers

Division of Signed Numbers

Rule 8. To divide two numbers, divide their absolute values (being careful to make sure you don't divide by 0) and then follow the rules for multiplication of signed numbers.

If 0 is the dividend, then the quotient is 0. For instance, $\frac{0}{5} = 0$. But, if 0 is the divisor, then the quotient is undefined. Thus, $\frac{5}{0} \neq 0$ and $\frac{5}{0} \neq 5$. $\frac{5}{0}$ has no answer because division by 0 is undefined!

Problem Find the quotient.

Division is commonly indicated by the fraction bar.

a. $\frac{-120}{-3}$

b. $\frac{-120}{3}$

c. $\frac{120}{-3}$

d. $\frac{-120}{0}$

e. $\frac{0}{30}$

Solution

a. $\frac{-120}{-3}$

Step 1. Divide 120 by 3.

$$\frac{120}{3} = 40$$

Step 2. The signs are the same (both negative), so keep the quotient positive.

$$\frac{-120}{-3} = 40$$

b. $\frac{-120}{3}$

Step 1. Divide 120 by 3.

$$\frac{120}{3} = 40$$

Step 2. The signs are opposites (one negative and one positive), so make the quotient negative.

$$\frac{-120}{3} = -40$$

c. $\frac{120}{-3}$

Step 1. Divide 120 by 3.

$$\frac{120}{3} = 40$$

Step 2. The signs are opposites (one positive and one negative), so make the quotient negative.

$$\frac{120}{-3} = -40$$

d. $\frac{-120}{0}$

Step 1. The divisor (denominator) is 0, so the quotient is undefined.

$$\frac{-120}{0} = \text{undefined}$$

e. $\frac{0}{30}$

Step 1. The dividend (numerator) is 0, so the quotient is 0.

$$\frac{0}{30} = 0$$

To be successful in arithmetical computation, you must memorize the rules for adding, subtracting, multiplying, and dividing signed numbers. Of course, when you do a computation, you don't have to write out all the steps. For instance, you can mentally ignore the signs to obtain the absolute values, do the necessary computation or computations, and then make sure your result has the correct sign.

Exercise 2

For 1–3, simplify.

1. $|-45|$
2. $|58|$
3. $|-5|$

For 4 and 5, state in words.

4. $-9+-(-4)=-9+4$

5. $-9-(-4)=-9+4$

For 6–17, compute as indicated.

6. $-80+-40$

7. $\frac{18}{-3}$

8. $(-100)(-8)$

9. $\frac{400}{2}$

10. $-458+0$

11. $4(-3)(0)(999)(-500)$

12. $\frac{0}{-56}$

13. $\frac{-1,400}{-700}$

14. $\frac{65}{-65}$

15. $\frac{40}{0}$

16. $(-3)(1)(-1)(-5)(-2)(2)(-10)$

17. $(-3)(1)(-1)(-5)(-2)(0)(-10)$

For 18–20, draw number line illustrations for the indicated sums.

18. $6+(-4)$

19. $-7+6$

20. $5+(-10)$

3

Exponents

In this chapter, you learn about exponents and exponentiation.

Exponential Notation

An *exponent* is a small raised number written to the upper right of a quantity, which is called the *base* for the exponent. For example, consider the product $3\times3\times3\times3\times3$, in which the same number is repeated as a factor multiple times. The exponential form for $3\times3\times3\times3\times3$ is 3^5. The number 3 is the *base*, and the small 5 to the upper right of 3 is the *exponent*.

The exponential form of a product of repeated factors is a shortened notation for the product.

Most commonly, you read 3^5 as either "three to the fifth" or "three raised to the fifth power." When the exponent is 1, as in 7^1, you say, "seven to the first." When 2 is the exponent, as in 6^2, you say, "six squared," and when 3 is the exponent, as in 5^3, you say, "five cubed."

Usually the exponent 1 is not written for expressions to the first power; that is, $7^1 = 7$.

Problem Express the exponential form in words.

a. 2^5

b. 5^6

c. 4^3

d. $(-13)^2$

e. 10^1

Solution

a. 2^5

Step 1. Identify the base.

The base is 2.

Step 2. Identify the exponent.

The exponent is 5.

Step 3. Express 2^5 in words.

2^5 is "two to the fifth."

b. 5^6

Step 1. Identify the base.

The base is 5.

Step 2. Identify the exponent.

The exponent is 6.

Step 3. Express 5^6 in words.

5^6 is "five to the sixth."

c. 4^3

Step 1. Identify the base.

The base is 4.

Step 2. Identify the exponent.

The exponent is 3.

Step 3. Express 4^3 in words.

4^3 is "four cubed."

d. $(-13)^2$

Step 1. Identify the base.

The base is -13.

Step 2. Identify the exponent.

The exponent is 2.

Step 3. Express $(-13)^2$ in words.

$(-13)^2$ is "negative thirteen squared."

When you use a negative number as the base in an exponential form, enclose the negative number in parentheses.

e. 10^1

Step 1. Identify the base.

The base is 10.

Step 2. Identify the exponent.

The exponent is 1.

Step 3. Express 10^1 in words.

10^1 is "ten to the first" or because $10^1 = 10$, simply "10."

Natural Number Exponents

When the exponent is a *natural number*, it tells you *how many times to use the base as a factor.*

Recall that the natural numbers are 1, 2, 3, and so on.

Problem Write the indicated product in exponential form.

a. $2 \times 2 \times 2 \times 2 \times 2 \times 2 \times 2$

b. $(-3)(-3)(-3)(-3)(-3)(-3)$

Solution

a. $2 \times 2 \times 2 \times 2 \times 2 \times 2 \times 2$

Step 1. Count how many times 2 is a factor.

$$\underbrace{2 \times 2 \times 2 \times 2 \times 2 \times 2 \times 2}_{\text{Seven factors of 2}}$$

Step 2. Write the indicated product as an exponential expression with 2 as the base and 7 as the exponent.

$$2 \times 2 \times 2 \times 2 \times 2 \times 2 \times 2 = 2^7$$

b. $(-3)(-3)(-3)(-3)(-3)(-3)$

Step 1. Count how many times -3 is a factor.

$$\underbrace{(-3)(-3)(-3)(-3)(-3)(-3)}_{\text{Six factors of } -3}$$

Step 2. Write the indicated product as an exponential expression with −3 as the base and 6 as the exponent.

$$(-3)(-3)(-3)(-3)(-3)(-3) = (-3)^6$$

In the above problem, you must enclose the −3 in parentheses to show that −3 is the number that is used as a factor six times.

To *evaluate* the exponential form 3^5, you do to the base what the exponent tells you to do. The result you get is the fifth *power* of 3, as shown in Figure 3.1.

Problem Evaluate 3^5.

Solution

Step 1. Write 3^5 in product form, using 3 as a factor five times.

$$3^5 = 3 \times 3 \times 3 \times 3 \times 3$$

Step 2. Do the multiplication.

$$3^5 = 3 \times 3 \times 3 \times 3 \times 3 = 243 \text{ (fifth power of 3)}$$

$3^5 \neq 3 \times 5$; $3^5 = 243$, but $3 \times 5 = 15$. Don't multiply the base by the exponent! That is a common mistake.

Note: To express $3^5 = 243$ in words, say either "three to the fifth is two hundred forty-three" or "three raised to the fifth power is two hundred forty-three."

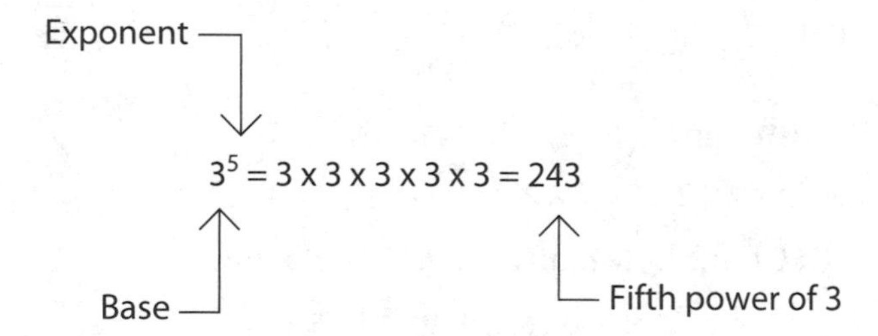

Figure 3.1 Parts of an exponential form

Problem Evaluate the expression.

a. 2^5

b. 5^6

c. 4^3

d. $(-13)^2$

e. 10^1

Solution

a. 2^5

Step 1. Write 2^5 in product form, using 2 as a factor five times.

$$2^5 = 2 \times 2 \times 2 \times 2 \times 2$$

Step 2. Do the multiplication.

$$2 \times 2 \times 2 \times 2 \times 2 = 32$$

Step 3. Review the main results.

$$2^5 = 2 \times 2 \times 2 \times 2 \times 2 = 32$$

b. 5^6

Step 1. Write 5^6 in product form, using 5 as a factor six times.

$$5^6 = 5 \times 5 \times 5 \times 5 \times 5 \times 5$$

Step 2. Do the multiplication.

$$5 \times 5 \times 5 \times 5 \times 5 \times 5 = 15{,}625$$

Step 3. Review the main results.

$$5^6 = 5 \times 5 \times 5 \times 5 \times 5 \times 5 = 15{,}625$$

c. 4^3

Step 1. Write 4^3 in product form, using 4 as a factor three times.

$$4^3 = 4 \times 4 \times 4$$

Step 2. Do the multiplication.

$$4 \times 4 \times 4 = 64$$

Step 3. Review the main results.

$$4^3 = 4 \times 4 \times 4 = 64$$

d. $(-13)^2$

Step 1. Write $(-13)^2$ in product form, using -13 as a factor two times.

$$(-13)^2 = (-13)(-13)$$

Step 2. Do the multiplication.

$$(-13)(-13) = 169$$

Step 3. Review the main results.

$$(-13)^2 = (-13)(-13) = 169$$

e. 10^1

Step 1. Write 10 as a factor one time.

$$10^1 = 10$$

You likely are most familiar with natural number exponents, but natural numbers are not the only numbers you can use as exponents. Here are two other types of exponents.

Zero Exponents

A *zero exponent* on a *nonzero* number tells you to *put 1 as the answer when you evaluate. Caution*: It's important to remember that when you use 0 as an exponent, the base *cannot* be 0. The expression 0^0 has no meaning. You say, "zero to the zero power is undefined."

Problem Evaluate.

a. 2^0

b. $(-25)^0$

c. 0^0

d. 100^0

e. 1^0

Solution

a. 2^0

Step 1. The exponent is 0, so put 1 as the answer.

$$2^0 = 1$$

$2^0 \neq 0$ and also $2^0 \neq 2$. A zero exponent gives 1 as the answer, so $2^0 = 1$.

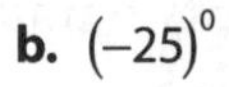

b. $(-25)^0$

Step 1. The exponent is 0, so put 1 as the answer.

$$(-25)^0 = 1$$

c. 0^0

Step 1. The exponent is 0, but 0^0 has no meaning, so put undefined as the answer.

0^0 is undefined.

d. 100^0

Step 1. The exponent is 0, so put 1 as the answer.

$100^0 = 1$

e. 1^0

Step 1. The exponent is 0, so put 1 as the answer.

$1^0 = 1$

Negative Exponents

A *negative exponent* on a *nonzero* number tells you to obtain the *reciprocal of the corresponding expression that has a positive exponent. Caution*: You *cannot* use 0 as a base for negative exponents. When you evaluate such expressions, you get 0 in the denominator, meaning that you have division by 0, which is undefined.

The reciprocal of a quantity is a fraction that has 1 in the numerator and the quantity in the denominator; for instance, the reciprocal of 2^5 is $\frac{1}{2^5}$.

Problem Evaluate.

a. 2^{-5}

b. 5^{-6}

c. 4^{-3}

d. $(-13)^{-2}$

e. 10^{-1}

f. 0^{-3}

Solution

a. 2^{-5}

Step 1. Write the expression using 5 instead of -5 as the exponent.

2^5

Step 2. Write the reciprocal of 2^5.

$$\frac{1}{2^5}$$

Step 3. Evaluate the denominator, 2^5.

$$\frac{1}{2^5} = \frac{1}{2 \times 2 \times 2 \times 2 \times 2} = \frac{1}{32}$$

Step 4. Review the main results.

$$2^{-5} = \frac{1}{2^5} = \frac{1}{32}$$

$2^{-5} \neq -\frac{1}{32}$. A negative exponent does not make a power negative. $2^{-5} = \frac{1}{32}$.

b. 5^{-6}

Step 1. Write the expression using 6 instead of –6 as the exponent.

5^6

Step 2. Write the reciprocal of 5^6.

$$\frac{1}{5^6}$$

Step 3. Evaluate the denominator, 5^6.

$$\frac{1}{5^6} = \frac{1}{5 \times 5 \times 5 \times 5 \times 5 \times 5} = \frac{1}{15{,}625}$$

Step 4. Review the main results.

$$5^{-6} = \frac{1}{5^6} = \frac{1}{15{,}625}$$

c. 4^{-3}

Step 1. Write the expression using 3 instead of –3 as the exponent.

4^3

Step 2. Write the reciprocal of 4^3.

$$\frac{1}{4^3}$$

Step 3. Evaluate the denominator, 4^3.

$$\frac{1}{4^3} = \frac{1}{4 \times 4 \times 4} = \frac{1}{64}$$

Step 4. Review the main results.

$$4^{-3} = \frac{1}{4^3} = \frac{1}{64}$$

d. $(-13)^{-2}$

Step 1. Write the expression using 2 instead of –2 as the exponent.

$$(-13)^2$$

Step 2. Write the reciprocal of $(-13)^2$.

$$\frac{1}{(-13)^2}$$

Step 3. Evaluate the denominator, $(-13)^2$.

$$\frac{1}{(-13)^2} = \frac{1}{(-13)(-13)} = \frac{1}{169}$$

Step 4. Review the main results.

$$(-13)^{-2} = \frac{1}{(-13)^2} = \frac{1}{169}$$

e. 10^{-1}

Step 1. Write the expression using 1 instead of –1 as the exponent.

$$10^1 = 10$$

Step 2. Write the reciprocal of 10.

$$\frac{1}{10}$$

Step 3. Review the main results.

$$10^{-1} = \frac{1}{10}$$

f. 0^{-3}

Step 1. The base is 0, so the expression is undefined.

0^{-3} is undefined.

Exercise 3

For 1–4, express the exponential form in words.

1. 6^5
2. $(-5)^4$
3. 4^0
4. $(-9)^2$

For 5 and 6, write the indicated product in exponential form.

5. $(-4)(-4)(-4)(-4)(-4)$
6. $8 \times 8 \times 8 \times 8 \times 8 \times 8 \times 8$

For 7–15, evaluate, if possible.

7. 2^8
8. 5^4
9. $(-4)^5$
10. 0^9
11. $(-2)^0$
12. 0^{-4}
13. 3^{-4}
14. $(-15)^{-2}$
15. 4^{-2}

4

Order of Operations

In this chapter, you apply your skills in computation to perform a series of indicated numerical operations. This chapter lays the foundation for numerical calculations by introducing you to the order of operations.

Grouping Symbols

Grouping symbols such as parentheses ( ), brackets [ ], and braces { } are used to keep things together that belong together. Fraction bars and absolute value bars are grouping symbols as well. When you do computations in numerical expressions, do operations in grouping symbols first. It is *very important* that you do so when you have addition or subtraction inside the grouping symbol.

Keep in mind, however, that parentheses also are used to indicate multiplication as in $(-5)(-8)$ or for clarity as in $-(-35)$.

Fraction bars indicate division. For instance, $\frac{18}{-2}$ means $18 \div -2$.

Grouping symbols say "Do me first!"

Problem Evaluate each expression.

a. $(1+1)^4$

b. $\frac{4+10}{4}$

c. $\frac{-7+25}{3-5}$

d. $|8+-15|$

Solution

a. $(1+1)^4$

Step 1. Parentheses are a grouping symbol, so do $1+1$ *first.*

$$(1+1)^4 = 2^4$$

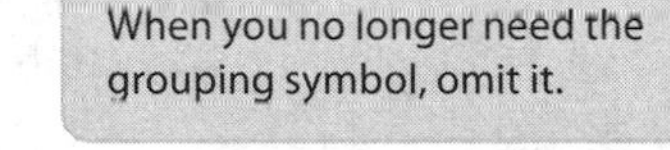

Step 2. Evaluate 2^4.

$$2^4 = 16$$

> $(1+1)^4 \neq 1^4 + 1^4$; $(1+1)^4 = 16$, but $1^4 + 1^4 = 1 + 1 = 2$. Not performing the addition, 1 + 1, inside the parentheses *first* can lead to an incorrect result.

Step 3. Review the main results.

$$(1+1)^4 = 2^4 = 16$$

b. $\dfrac{4+10}{4}$

Step 1. The fraction bar is a grouping symbol, so do the addition, $4+10$, over the fraction bar *first.*

$$\frac{4+10}{4} = \frac{14}{4}$$

Step 2. Simplify $\dfrac{14}{4}$.

$$= \frac{7}{2}$$

> $\dfrac{4+10}{4} \neq \dfrac{\cancel{4}+10}{\cancel{4}} \neq \dfrac{10}{1}$. Not performing the addition, $4+10$, *first* can lead to an incorrect result.

Step 3. Review the main results.

$$\frac{4+10}{4} = \frac{14}{4} = \frac{7}{2}$$

c. $\dfrac{-7+25}{3-5}$

Step 1. The fraction bar is a grouping symbol, so do the addition, $-7+25$, over the fraction bar and the subtraction, $3-5$, under the fraction bar *first.*

$$\frac{-7+25}{3-5} = \frac{18}{-2}$$

Step 2. Compute $\frac{18}{-2}$.

$$\frac{18}{-2} = 18 \div -2 = -9$$

Step 3. Review the main results.

$$\frac{-7+25}{3-5} = \frac{18}{-2} = -9$$

$\frac{-7+25}{3-5} \neq \frac{-7+5}{3-1}$; $\frac{-7+25}{3-5} = -9$, but $\frac{-7+5}{3-1} = \frac{-2}{2} = -1$. Not performing the addition, $-7+25$, and the subtraction, $3-5$, *first* can result in a cancellation error.

d. $|8+-15|$

Step 1. Absolute value bars are a grouping symbol, so do $8+-15$ *first*.

$$|8+-15| = |-7|$$

Step 2. Evaluate $|-7|$.

$$= 7$$

Step 3. Review the main results.

$$|8+-15| = |-7| = 7$$

$|8+-15| \neq |8|+|-15|$; $|8+-15| = 7$, but $|8|+|-15| = 8+15 = 23$. Not performing the addition, $8+-15$, *first* can lead to an incorrect result.

PEMDAS

Mathematical expressions like $60 \div 12 - 3 \times 4 + (1+1)^3$ and $\frac{-7+25}{3-5}$ are *numerical expressions*. When you evaluate numerical expressions by doing the indicated math, you must follow the order of operations. Use the mnemonic "**P**lease **E**xcuse **M**y **D**ear **A**unt **S**ally"—abbreviated as PE(MD)(AS)—to help you remember the following order.

Order of Operations

1. Do computations inside **P**arentheses (or other grouping symbols).
2. Do **E**xponents (also, evaluate absolute values and roots).
3. Do **M**ultiplication and **D**ivision, in the order in which they occur from left to right.
4. Do **A**ddition and **S**ubtraction, in the order in which they occur from left to right.

In the order of operations, multiplication does not always have to be done before division, or addition before subtraction. You multiply and divide in the order in which they occur in the problem. Similarly, you add and subtract in the order in which they occur in the problem.

Problem Evaluate the expression.

a. $60 \div 12 - 3 \times 4 + (1+1)^3$

b. $100 + 8 \times 3^2 - 63 \div (2+5)$

c. $\dfrac{-7+25}{3-5} + |8 + -15| - (5-3)^3$

d. $(12-5) - (5-12)$

e. $-8 + 2(-1)^2 + 6$

f. $2 + 10 \times 3 + 1$

g. $(2 \times 5)^2 + 2 \times 5^2$

h. $2 + 5(6-4)$

Solution

a. $60 \div 12 - 3 \times 4 + (1+1)^3$

Step 1. Do 1 + 1 inside the parentheses.

$= 60 \div 12 - 3 \times 4 + 2^3$

Step 2. Do the exponent: 2^3.

$= 60 \div 12 - 3 \times 4 + 8$

Step 3. Do multiplication and division, from left to right.

Do $60 \div 12$ because division occurs first.

$= 5 - 3 \times 4 + 8$

Do 3×4.

$= 5 - 12 + 8$

> $5 - 3 \times 4 + 8 \neq 2 \times 12$. Multiply *before* adding or subtracting—when no grouping symbols are present.

Step 4. Do addition and subtraction, from left to right.

Do $5 - 12$ because subtraction occurs first.

$= -7 + 8$

Do $-7 + 8$.

$= 1$

Step 5. Review the main results.

$60 \div 12 - 3 \times 4 + (1+1)^3$

$= 60 \div 12 - 3 \times 4 + 2^3$

$= 60 \div 12 - 3 \times 4 + 8$

$= 5 - 12 + 8$

$= 1$

b. $100 + 8 \times 3^2 - 63 \div (2+5)$

Step 1. Compute 2 + 5 inside the parentheses.

$= 100 + 8 \times 3^2 - 63 \div 7$

Step 2. Do the exponent: 3^2.

$= 100 + 8 \times 9 - 63 \div 7$

$8 \times 3^2 \neq 24^2$; $8 \times 3^2 = 8 \times 9 = 72$, but $24^2 = 576$. Do exponents *before* multiplication.

Step 3. Do multiplication and division, from left to right.

Do 8×9 because multiplication occurs first.

$= 100 + 72 - 63 \div 7$

$100 + 8 \times 9 \neq 108 \times 9$. Do multiplication *before* addition (except when a grouping symbol indicates otherwise).

Do $63 \div 7$.

$= 100 + 72 - 9$

$72 - 63 \div 7 \neq 9 \div 7$. Do division *before* subtraction (except when a grouping symbol indicates otherwise).

Step 4. Do addition and subtraction, from left to right.

Do $100 + 72$ because addition occurs first.

$= 172 - 9$

Do $172 - 9$.

$= 163$

Step 5. Review the main results.

$100 + 8 \times 3^2 - 63 \div (2+5)$

$= 100 + 8 \times 3^2 - 63 \div 7$

$= 100 + 8 \times 9 - 63 \div 7$

$= 100 + 72 - 9$

$= 163$

c. $\frac{-7+25}{3-5}+|8+-15|-(5-3)^3$

Step 1. Compute quantities in grouping symbols.

$=\frac{18}{-2}+|-7|-2^3$

Step 2. Evaluate $|-7|$ and 2^3.

$=\frac{18}{-2}+7-8$

Evaluate absolute value expressions and exponents *before* multiplication or division.

Step 3. Do division: $\frac{18}{-2}$.

$=-9+7-8$

Step 4. Do addition and subtraction, from left to right.

Do $-9+7$ because addition occurs first.

$=-2-8$

Do $-2-8$.

$=-10$

Step 5. Review the main results.

$\frac{-7+25}{3-5}+|8+-15|-(5-3)^3$

$=\frac{18}{-2}+|-7|-2^3$

$=\frac{18}{-2}+7-8$

$=-9+7-8$

$=-10$

d. $(12-5)-(5-12)$

Step 1. Compute quantities in parentheses.

$=7-(-7)$

Step 2. Do subtraction: $7-(-7)$.

$=7+7=14$

Step 3. Review the main results.

$$(12-5)-(5-12)$$
$$=7-(-7)$$
$$=7+7$$
$$=14$$

e. $-8+2(-1)^2+6$

Step 1. Do the exponent: $(-1)^2$.

$$=-8+2\cdot 1+6$$

Step 2. Do multiplication: $2\cdot 1$.

$2\cdot 1=2\times 1$. You can use a raised dot to show multiplication.

$$=-8+2+6$$

Step 3. Do addition.

$$=-8+2+6=-6+6=0$$

Step 4. Review the main results.

$$-8+2(-1)^2+6$$
$$=-8+2\cdot 1+6$$
$$=-8+2+6$$
$$=0$$

f. $2+10\times 3+1$

Step 1. Do multiplication: 10×3.

$$=2+30+1$$

Step 2. Do addition.

$$=2+30+1=32+1=33$$

Step 3. Review the main results.

$$2+10\times 3+1$$
$$=2+30+1$$
$$=33$$

$2+10\times 3+1\neq 12\times 4$; $2+10\times 3+1=33$, but $12\times 4=48$. Don't add (or subtract) before multiplying, unless you have a grouping symbol that tells you to do so.

g. $(2\times5)^2+2\times5^2$

Step 1. Do 2×5 inside the parentheses.

$=10^2+2\times5^2$

Step 2. Do the exponents.

$=100+2\times25$

Step 3. Do multiplication: 2×25.

$=100+50$

$2\times5^2\neq10^2$; $2\times5^2=2\times25=50$, but $10^2=100$.

Step 4. Do addition.

$=150$

Step 5. Review the main results.

$(2\times5)^2+2\times5^2$

$=10^2+2\times5^2$

$=100+2\times25$

$=100+50$

$=150$

h. $2+5(6-4)$

Step 1. Do parentheses.

$=2+5(2)$

$2+5(6-4)\neq7(6-4)$ because parentheses and multiplication come before addition.

Step 2. Do multiplication.

$=2+10$

$5(2)=5\times2$. You can use parentheses to show multiplication.

Step 3. Do addition.

$=12$

Step 4. Review the main results.

$2+5(6-4)$

$=2+5(2)$

$=2+10$

$=12$

Exercise 4

Evaluate the expression.

1. $(5+7)6-10$
2. $(-7^2)(6-8)$
3. $(2-3)(-20)$
4. $3(-2)-\frac{10}{-5}$
5. $9-\frac{20+22}{6}-2^3$
6. $-2^2\cdot-3-(15-4)^2$
7. $5(11-3-6\cdot 2)^2$
8. $-10-\frac{-8-(3\cdot-3+15)}{2}$
9. $\frac{7^2-8\cdot 5+3^4}{3\cdot 2-36\div 12}$
10. $\frac{5\ |\ 5|}{20^2}$
11. $\frac{3}{2}\left(-\frac{2}{3}\right)-\frac{1}{4}(-5)+\frac{15}{7}\left(-\frac{7}{3}\right)$

5

Fractions

In this chapter, you learn how to work with fractions.

Fraction Concepts

As greater precision in measurement was needed, the concept of fractions (rational numbers) was invented. If a *unit* is broken into equal parts, then a fraction represents one or more of those equal parts. For example, if an inch is broken into 10 equal parts, then $\frac{7}{10}$ represents seven of those equal parts. When the *unit* (the distance between 0 and 1) is divided into 10 equal parts, the fraction $\frac{7}{10}$ is a point on the number line, as shown in Figure 5.1.

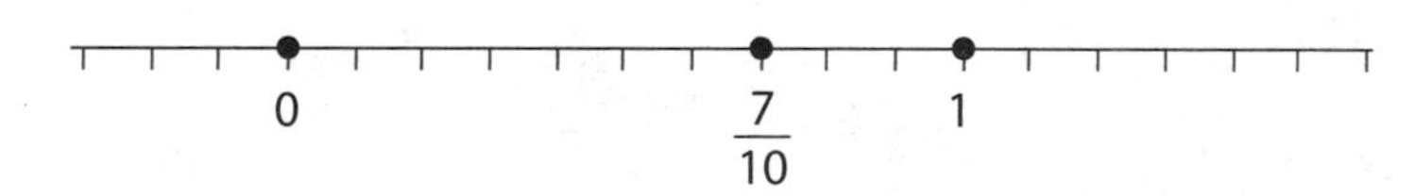

Figure 5.1 The fraction $\frac{7}{10}$

The division line in a fraction is the *fraction bar*. The number above the fraction bar is the *numerator*, and the number below the fraction bar is the *denominator*. For example, in the fraction $\frac{7}{10}$, 7 is the numerator, and 10 is the denominator.

The denominator of a fraction cannot be 0.

Reducing Fractions to Lowest Form

A fundamental principle of fractions is the *cancellation law.*

Cancellation Law

If both the numerator and the denominator of a fraction are multiplied by the same nonzero number, then the value of the fraction is unchanged. That is, $\frac{a \cdot n}{b \cdot n} = \frac{a}{b}$ provided that $b \neq 0$ and/or $n \neq 0$.

The cancellation law is not valid if the operation is addition instead of multiplication. That is, $\frac{a+n}{b+n} \neq \frac{a}{b}$. This is a common error observed in the arithmetic of fractions.

A fraction is in *lowest form* when all the common factors of the numerator and denominator have been cancelled. The process of cancelling the common factors is called *reducing the fraction.*

Problem Reduce $\frac{30}{36}$ to lowest form.

Solution

Step 1. Write the numerator and denominator in a factored form.

$$\frac{30}{36} = \frac{15 \cdot 2}{18 \cdot 2}$$

Step 2. Cancel the common factor.

$$= \frac{15 \cdot 2}{18 \cdot 2} = \frac{15 \cdot \cancel{2}}{18 \cdot \cancel{2}} = \frac{15}{18}$$

Step 3. If possible, factor again.

$$\frac{15}{18} = \frac{5 \cdot 3}{6 \cdot 3}$$

Step 4. Cancel the common factor.

$$= \frac{5 \cdot 3}{6 \cdot 3} = \frac{5 \cdot \cancel{3}}{6 \cdot \cancel{3}} = \frac{5}{6}$$

Step 5. If there are no more common factors, state the lowest form.

$$\frac{30}{36} = \frac{5}{6}$$

You can simplify the reduction process considerably if you factor by using the *greatest common factor* (GCF). As implied by its name, the GCF of two numbers is the largest factor that is common to the two numbers. For example, the GCF of

30 and 36 is 6. The factoring in the previous problem could have been $\frac{30}{36}=\frac{5\cdot 6}{6\cdot 6}$, and then the 6 canceled to get the reduced form of $\frac{5}{6}$ in just two steps. You can determine the GCF of two numbers by listing all the factors of the two numbers and selecting the largest factor that is common to both.

Problem Find the GCF of 30 and 36.

Solution

Step 1. List all the factors of 30 and all the factors of 36.

The factors of 30 are 1, 2, 3, 5, 6, 10, 15, 30, and the factors of 36 are 1, 2, 3, 4, 6, 9, 12, 18, 36.

Step 2. Examine the two lists and select the greatest factor common to both.

6 is the greatest factor that is common to both lists. Thus, the GCF of 30 and 36 is 6.

Another way to accomplish the reduction process is to *divide* the numerator and denominator by the same number and write the resulting quotients as the equivalent fraction. The common way of writing the reduction is

$\frac{30\div 6}{36\div 6}=\frac{\overset{5}{\cancel{30}}}{\underset{6}{\cancel{36}}}=\frac{5}{6}$, where you divide both the numerator and the denominator by 6.

Problem Write $\frac{66}{88}$ in lowest form.

Solution

Step 1. Divide the numerator and denominator by 22.

$$\frac{66\div 22}{88\div 22}=\frac{\overset{3}{\cancel{66}}}{\underset{4}{\cancel{88}}}=\frac{3}{4}$$

Step 2. If there are no more common factors, state the lowest form.

$$\frac{66}{88}=\frac{3}{4}$$

The process of reduction is very useful in operations involving fractions, especially in the area of algebra. Also, $\frac{3}{4}$ is much easier to locate or visualize on the number line than is $\frac{66}{88}$. Reduction is a simple way of writing a fraction in a more recognizable form to help determine its relative size.

Equivalent Fractions

Fractions are the only real numbers that use two numerical components (a numerator and a denominator) to express the number. Consequently, a concept known as *equivalent fractions* is peculiar to fractions. There are several ways to express this idea.

Equivalent Fractions

Two fractions, $\frac{a}{b}$ and $\frac{c}{d}$, are *equivalent* if and only if

(a) they locate the same point on the number line;

(b) one can be reduced to the other; or

(c) $ad = bc$.

Each of the ways of expressing equivalence has its use, but the method in (c) above is of special significance when doing arithmetic with fractions.

Adding and Subtracting Fractions

Addition and Subtraction of Fractions

Like denominators: If $\frac{a}{d}$ and $\frac{c}{d}$ are fractions with *like* denominators, then

$\frac{a}{d}+\frac{c}{d}=\frac{a+c}{d}$ and $\frac{a}{d}-\frac{c}{d}=\frac{a-c}{d}$

Unlike denominators: If $\frac{a}{b}$ and $\frac{c}{d}$ are any two fractions, then

$\frac{a}{b}+\frac{c}{d}=\frac{ad+bc}{bd}$ and $\frac{a}{b}-\frac{c}{d}=\frac{ad-bc}{bd}$

Problem Perform the indicated operation.

a. $\frac{7}{15}+\frac{11}{15}$

b. $\frac{17}{31}-\frac{22}{15}$

c. $\frac{4}{5}+\frac{6}{11}$

d. $\frac{4}{5}-\frac{6}{11}$

Solution

a. $\frac{7}{15}+\frac{11}{15}$

Step 1. The denominators are alike, so apply $\frac{a}{d}+\frac{c}{d}=\frac{a+c}{d}$ and reduce, if needed.

$$\frac{7}{15}+\frac{11}{15}$$

$$=\frac{7+11}{15}=\frac{18}{15}=\frac{18\div 3}{15\div 3}=\frac{6}{5}$$

b. $\frac{17}{31}-\frac{22}{15}$

Step 1. The denominators are not alike, so apply $\frac{a}{b}-\frac{c}{d}=\frac{ad-bc}{bd}$ and reduce, if needed.

$$\frac{17}{31}-\frac{22}{15}$$

$$=\frac{17\cdot 15-31\cdot 22}{31\cdot 15}=\frac{255-682}{465}=\frac{-427}{465}=-\frac{427}{465}$$

c. $\frac{4}{5}+\frac{6}{11}$

Step 1. The denominators are not alike, so apply $\frac{a}{b}+\frac{c}{d}=\frac{ad+bc}{bd}$ and reduce, if needed.

$$\frac{4}{5}+\frac{6}{11}$$

$$=\frac{4\cdot 11+5\cdot 6}{5\cdot 11}=\frac{44+30}{55}=\frac{74}{55}$$

d. $\frac{4}{5}-\frac{6}{11}$

Step 1. The denominators are not alike, so apply $\frac{a}{b}-\frac{c}{d}=\frac{ad-bc}{bd}$ and reduce, if needed.

$$\frac{4}{5}-\frac{6}{11}$$

$$=\frac{4\cdot 11-5\cdot 6}{5\cdot 11}=\frac{44-30}{55}=\frac{14}{55}$$

Because adding or subtracting fractions is very simple when they have the same denominator, a third method of combining any two fractions is to write the two fractions as equivalent fractions with like denominators and then add (or subtract). This method, referred to as *finding a common denominator*, is used often.

For example, $\frac{3}{4}=\frac{3\cdot 7}{4\cdot 7}=\frac{21}{28}$ and $\frac{5}{7}=\frac{5\cdot 4}{7\cdot 4}=\frac{20}{28}$. Hence, the cancellation law ensures both can be written as equivalent fractions with the same denominator. This is not a unique process because $\frac{3}{4}=\frac{3\cdot 28}{4\cdot 28}=\frac{84}{112}$ and $\frac{5}{7}=\frac{5\cdot 16}{7\cdot 16}=\frac{80}{112}$. Again, both can be written as equivalent fractions with like denominators. However, 28 is the least common denominator and is preferred in most cases.

Here is an example of finding a common denominator to subtract two fractions.

Problem Compute as indicated: $\frac{4}{5}-\frac{6}{11}$.

Solution

Step 1. Find the common denominator.

The common denominator of 5 and 11 is 55.

Step 2. Write $\frac{4}{5}$ and $\frac{6}{11}$ as equivalent fractions that have the denominator 55.

$$\frac{4\cdot 11}{5\cdot 11}=\frac{44}{55} \text{ and } \frac{6\cdot 5}{11\cdot 5}=\frac{30}{55}$$

Step 3. Subtract the equivalent fractions and reduce, if needed.

$$\frac{44}{55}-\frac{30}{55}=\frac{14}{55}$$

The method of combining fractions by finding a common denominator is used by many, but using $\frac{a}{b}+\frac{c}{d}=\frac{ad+bc}{bd}$ or $\frac{a}{b}-\frac{c}{d}=\frac{ad-bc}{bd}$ as shown above for combining two fractions can be faster; it also automatically produces a common denominator and transfers easily to algebra. Nevertheless, you should choose the method that works best for you.

Multiplying and Dividing Fractions

Multiplication of Fractions

If $\frac{a}{b}$ and $\frac{c}{d}$ are two fractions, then $\frac{a}{b}\cdot\frac{c}{d}=\frac{a\cdot c}{b\cdot d}$.

To multiply two fractions, multiply the two numerators and the two denominators and then reduce, if needed.

Problem Multiply: $\frac{3}{11}\cdot\frac{5}{6}$.

Solution

Step 1. Apply $\frac{a}{b}\cdot\frac{c}{d}=\frac{a\cdot c}{b\cdot d}$ and reduce, if needed.

$$\frac{3}{11}\cdot\frac{5}{6}$$

$$=\frac{3\cdot 5}{11\cdot 6}=\frac{15}{66}=\frac{15\div 3}{66\div 3}=\frac{5}{22}$$

Common denominators are not needed when multiplying fractions.

The fraction $\frac{d}{c}$ is the reciprocal of the fraction $\frac{c}{d}$ provided $\frac{c}{d}\neq 0$. This concept is used when you divide fractions.

Division of Fractions

If $\frac{a}{b}$ and $\frac{c}{d}$ are two fractions and $\frac{c}{d}\neq 0$, then

$\frac{a}{b}\div\frac{c}{d}=\frac{a}{b}\cdot\frac{d}{c}=\frac{a\cdot d}{b\cdot c}$.

To divide two fractions, multiply the first fraction by the reciprocal of the divisor fraction.

Problem Divide: $\frac{27}{12} \div \frac{3}{4}$.

Solution

Step 1. Apply $\frac{a}{b} \div \frac{c}{d} = \frac{a}{b} \cdot \frac{d}{c} = \frac{a \cdot d}{b \cdot c}$ and reduce, if needed.

$$\frac{27}{12} \div \frac{3}{4}$$

$$= \frac{27}{12} \cdot \frac{4}{3} = \frac{108}{36} = \frac{3 \cdot \cancel{36}}{1 \cdot \cancel{36}} = \frac{3}{1} = 3$$

Working with Mixed Numbers and Improper Fractions

A *mixed number* is a whole number in combination with a fraction such as $2\frac{5}{7}$ (read as "two and five-sevenths"). This mixed number is an alternate representation of the *improper fraction* $\frac{2 \cdot 7 + 5}{7} = \frac{19}{7}$. Any mixed number can be converted to its fractional form by the method $a\frac{c}{b} = \frac{a \cdot b + c}{b}$. Also, the added form is $2\frac{5}{7} = 2 + \frac{5}{7}$.

One way to add or subtract with mixed numbers is to change the mixed number to an improper fraction before adding or subtracting, as shown in the following problem.

Problem Compute as indicated. Write the answer as a mixed number.

a. Add $\frac{5}{6}$ and $3\frac{5}{7}$.

b. Subtract $3\frac{5}{7}$ from $5\frac{3}{5}$.

Solution

a. Add $\frac{5}{6}$ and $3\frac{5}{7}$.

Step 1. Convert the mixed number to an improper fraction.

$$3\frac{5}{7} = \frac{3 \cdot 7 + 5}{7} = \frac{26}{7}$$

Step 2. Add and reduce to lowest form, if needed.

$$\frac{5}{6}+\frac{26}{7}$$

$$=\frac{5\cdot7+6\cdot26}{6\cdot7}=\frac{35+156}{42}=\frac{191}{42}$$

Step 3. Write as a mixed number.

$$\frac{191}{42}=\frac{4\cdot42+23}{42}=4\frac{23}{42}$$

b. Subtract $3\frac{5}{7}$ from $5\frac{3}{5}$.

Step 1. Write both fractions as improper fractions.

$$3\frac{5}{7}=\frac{3\cdot7+5}{7}=\frac{26}{7} \text{ and } 5\frac{3}{5}=\frac{5\cdot5+3}{5}=\frac{28}{5}$$

Step 2. Subtract and reduce to lowest form, if needed.

$$\frac{28}{5}-\frac{26}{7}=\frac{28\cdot7-5\cdot26}{5\cdot7}=\frac{196-130}{35}=\frac{66}{35}$$

Step 3. Write as a mixed number.

$$\frac{66}{35}=\frac{1\cdot35+31}{35}=1\frac{31}{35}$$

Another way to add or subtract with mixed numbers is to work with the whole number parts and fractional parts separately, as shown in the following problem.

Problem Add $5\frac{3}{5}$ and $6\frac{4}{7}$.

Solution

Step 1. Write both fractions in added form.

$$5+\frac{3}{5} \text{ and } 6+\frac{4}{7}$$

Step 2. Add the whole numbers and then the fractions.

$$5+\frac{3}{5}+6+\frac{4}{7}$$
$$=11+\frac{3}{5}+\frac{4}{7}$$
$$=11+\frac{21+20}{35}$$
$$=11+\frac{41}{35}$$
$$=11+\frac{1\cdot 35+6}{35}$$
$$=11+1+\frac{6}{35}$$
$$=12+\frac{6}{35}=12\frac{6}{35}$$

When you multiply or divide mixed numbers, change mixed numbers to improper fractions before multiplying or dividing.

Problem Compute as indicated.

a. Divide $3\frac{3}{4}$ by $6\frac{3}{7}$.

b. Multiply $1\frac{1}{2}$ by $1\frac{1}{2}$.

Solution

a. Divide $3\frac{3}{4}$ by $6\frac{3}{7}$.

Step 1. Write both fractions as improper fractions.

$$3\frac{3}{4}=\frac{15}{4} \text{ and } 6\frac{3}{7}=\frac{45}{7}$$

Step 2. Do the division and reduce to lowest form, if needed.

$$3\frac{3}{4}\div 6\frac{3}{7}$$

$$= \frac{15}{4} \div \frac{45}{7} = \frac{15}{4} \cdot \frac{7}{45}$$

$$= \frac{\overset{1}{\cancel{15}}}{4} \cdot \frac{7}{\underset{3}{\cancel{45}}} = \frac{1 \cdot 7}{4 \cdot 3} = \frac{7}{12}$$

Note: You can divide out common factors before multiplying.

b. Multiply $1\frac{1}{2}$ by $1\frac{1}{2}$.

Step 1. Write both fractions as improper fractions.

$1\frac{1}{2} = \frac{3}{2}$ and $1\frac{1}{2} = \frac{3}{2}$

Step 2. Do the multiplication and reduce to lowest form, if needed.

$$1\frac{1}{2} \cdot 1\frac{1}{2}$$

$$\frac{3}{2} \cdot \frac{3}{2} = \frac{9}{4} = 2\frac{1}{4}$$

Exercise 5

Compute as indicated. Reduce to lowest form and write as a mixed number, when possible.

1. $\frac{3}{8} + \frac{7}{16}$
2. $\frac{7}{11} - \frac{4}{7}$
3. $2\frac{3}{4} + \frac{11}{12}$
4. $\frac{6}{5} \cdot \frac{15}{36}$
5. $\frac{8}{9} \div \frac{4}{7}$
6. $\frac{3}{8} - \frac{7}{24}$
7. $\left(\frac{6}{5} + \frac{3}{5}\right) \cdot \frac{5}{11}$
8. $\frac{7}{12} - \frac{4}{3} + \frac{5}{6}$
9. $\frac{5}{11} \div \frac{5}{7}$
10. $\frac{7}{11} \div \frac{8}{11}$
11. $3\frac{3}{5} - 2\frac{4}{7}$
12. $3\frac{2}{3} \div 4\frac{3}{5}$

6

Decimals

In this chapter, you learn how to work with decimals.

Decimal Concepts

Decimal fractions are fractions with a denominator that is some positive power of 10, such as $\frac{1}{10}, \frac{9}{100}, \frac{3}{1{,}000}$, and so forth. To represent these numbers, you extend the *place-value system* of numbers and use a decimal point to separate whole numbers from decimal fractions. The number 428.36 is a mixed decimal and means $(4 \cdot 100)+(2 \cdot 10)+(8 \cdot 1)+\left(3 \cdot \frac{1}{10}\right)+\left(6 \cdot \frac{1}{100}\right)$. You read 428.36 as "Four hundred twenty-eight and thirty-six hundredths" or as "Four hundred twenty-eight point thirty-six."

> Don't say "and" when reading whole numbers. For instance, 203 is "two hundred three," not "two hundred and three."

Hereinafter, mixed decimals and decimal fractions will be called simply decimals.

A place-value diagram for some of the positional values of the decimal system is shown in Figure 6.1.

The use of the decimal point is a very convenient way to represent decimal fractions. For instance, using the decimal point, you write $\frac{3}{10}+\frac{6}{100}+\frac{5}{1{,}000}$ as 0.365.

> The 0 in front of the decimal point is used as a way to make the decimal point noticeable for decimal fractions that are less than 1.

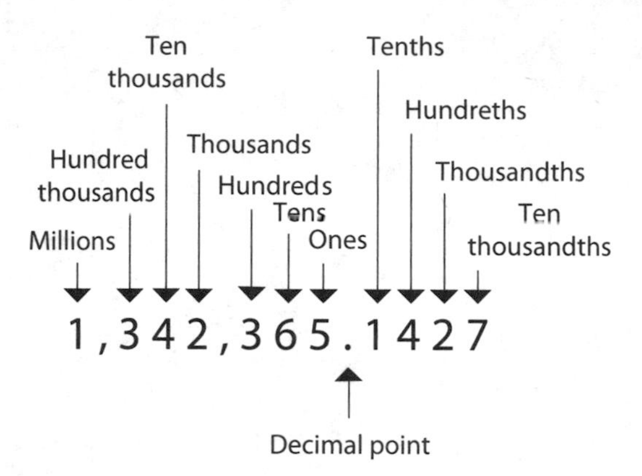

You read a decimal point as either "and" or "point." For example, 35.6 is read "Thirty-five and six-tenths" or as "Thirty-five point six."

Figure 6.1 Place values in the decimal system

You deal with the decimal point in arithmetic calculations by following some simple "rules" that are just mathematical shortcuts to ensure accuracy of the calculations.

Adding and Subtracting Decimals

Adding and subtracting are only done for like amounts. A statement such as "4 inches plus 7 inches" is meaningful, while "7 apples plus 5 inches" is meaningless. Similarly, with decimals, you combine tenths with tenths, hundredths with hundredths, and so on. Thus, when you add or subtract decimals, keep the decimal points lined up in the computation so that you are adding digits of like place values.

Problem Add 25.78, 241.342, and 12.5.

Solution

Step 1. Write the numbers in an addition column, being sure to line up the decimal points.

$$\begin{array}{r} 25.780 \\ 241.342 \\ +\ 12.500 \\ \hline \end{array}$$

Zeros are inserted where needed to make sure that all place values are in all the numbers.

Step 2. Add as with whole numbers.

$$\begin{array}{r} 25.780 \\ 241.342 \\ +\ 12.500 \\ \hline 279.622 \end{array}$$

Problem Subtract 168.274 from 6,547.34.

Solution

Step 1. Write the numbers in a subtraction column, being sure to line up the decimal points.

$$\begin{array}{r} 6{,}547.340 \\ -\ 168.274 \\ \hline \end{array}$$

When adding or subtracting decimals, you should insert zeros for missing place values.

Step 2. Subtract as you would with whole numbers.

$$\begin{array}{r} 6{,}547.340 \\ -\ 168.274 \\ \hline 6{,}379.066 \end{array}$$

The carrying and borrowing procedure is the same as with whole numbers.

Multiplying Decimals

A simple rule for multiplying two decimals is to sum the number of decimal places in the *multiplicand* (first factor) and the *multiplier* (second factor) and put this number of places in the product.

Problem Multiply 47.63 by 32.57.

Solution

Step 1. Write the numbers in a multiplication column.

$$\begin{array}{r} 47.63 \\ \times 32.57 \\ \hline \end{array}$$

You do not have to line up the decimal points for a multiplication problem.

Step 2. Perform the multiplication as you would with whole numbers.

$$\begin{array}{r} 47.63 \\ \times 32.57 \\ \hline 3341 \\ 238150 \\ 952600 \\ 14289000 \\ \hline 15513091 \end{array}$$

Step 3. Sum the number of decimal places in both the multiplicand and the multiplier, which is four in this problem, and put that number of decimal places in the final product.

$$\begin{array}{r} 47.63 \\ \times 32.57 \\ \hline 3341 \\ 238150 \\ 952600 \\ 14289000 \\ \hline 1{,}551.3091 \end{array}$$

Dividing Decimals

To devise a rule for dividing decimals, you employ the use of the cancellation law of fractions. For example, $\frac{6.25}{2.5} = \frac{(6.25)\cdot 10}{(2.5)\cdot 10} = \frac{62.5}{25}$. The technique, then, in dividing decimals, is to multiply both *dividend* (numerator) and *divisor* (denominator) by the appropriate power of 10 that will make the *divisor a whole number.* In practice, this strategy amounts to moving the decimal point *to the right* the appropriate number of places to make the divisor a whole number. Of course, the decimal point in the dividend must be moved in the same manner. Once this is done, the decimal point in the dividend and the *quotient* (answer) must be aligned.

Problem Divide 6.25 by 2.5.

Solution

Step 1. Write as a long division problem.

$2.5\overline{)6.25}$

Step 2. Move the decimal point one place to the right in both numbers to make the divisor a whole number.

$25\overline{)62.5}$

Step 3. Divide as you would whole numbers, keeping the decimal points aligned but ignoring the decimal point in the intermediate multiplications.

$$\begin{array}{r} 2.5 \\ 25\overline{)62.5} \\ \underline{50} \\ 125 \\ \underline{125} \\ 0 \end{array}$$

Step 4. State the main result.

$$\frac{6.25}{2.5} = 2.5$$

Step 5. Check the answer.

$$\begin{array}{r} 2.5 \\ \times\, 2.5 \\ \hline 125 \\ 500 \\ \hline 6.25 \end{array}$$

Calculators will automatically place the decimal point, but you should know these rules to check for errors that may occur in entering data into your calculator.

Rounding Decimals

Working with decimals can sometimes result in lengthy decimal expressions. In application, you may be interested in the decimal expression to only a few places. In this case, you use the technique of rounding to determine the final approximation. For example, if you want the decimal to only two places, you look at the digit in the third place to the right of the decimal point. If the third digit is 5 or greater, increase the digit in the second place by 1 and drop all digits past the second digit to the right of the decimal point. If the third digit is less than 5, leave the digit in the second place as is and drop all digits past the second digit to the right of the decimal point. For example, 45.57689 rounded to two places is 45.58. The process for all places is the same.

Problem Use a calculator to compute $\frac{67.893}{42.568}$ and then round the answer to two decimal places.

$\frac{67.893}{42.568} = 67.893 \div 42.568$. Recall from previous chapters that the fraction bar indicates division. It is important to recognize this use of the fraction bar.

Solution

Step 1. Do the division by calculator.

$$\frac{67.893}{42.568} \approx 1.594930464$$

Step 2. Check whether the digit in the third place to the right of the decimal point is 5 or greater or less than 5.

The digit in the third place is 4, which is less than 5.

Step 3. Leave the digit in the second place as is and drop all digits past the second digit to the right of the decimal point.

$$\frac{67.893}{42.568} \approx 1.59$$

Handheld calculators have taken the tedium out of arithmetic calculations and for that all are glad. Do the following problems by hand or with a calculator, but before you do them, predict the number of decimal places in the answer or, in the case of division, where the decimal point will be located.

Exercise 6

1. Add 45.716 and 3.92.
2. Subtract 1.8264 from 23.3728.
3. Multiply 0.214 by 1.93.
4. Multiply 1.21 by 0.0056.
5. Divide 0.1547 by 0.014.
6. Divide 2.916 by 0.36.
7. Divide 2.917 by 0.37 and round to three places.
8. Multiply 6.678 by 0.37 and round to two places.
9. Divide 3.977 by 0.0372 and round to three places.
10. Multiply 45.67892 by 0.0374583 and round to four places.

7

Percents

In this chapter, you learn about percents and their relationship to fractions and decimals.

Percent Concepts

A *percent* is a special type of fraction with a denominator of 100. The word *percent* means "per hundred." There are three ways of writing a percent: as a fraction, as a decimal, or with the special percent (%) symbol. For example, $\frac{3}{100} = 0.03 = 3\%$. The % symbol is a shorthand way of telling you to multiply the attached number by either $\frac{1}{100}$ or 0.01. You use the % symbol in writing, but calculations usually require percents to be in decimal or fraction form.

The word *rate* is sometimes used for *percent*.

To remember that % refers to $\frac{1}{100}$, think of the slash in the % symbol as a fraction bar and the two zeros as part of the number 100.

Changing Percents to Decimal Form

You can change a percent to decimal form by substituting 0.01 for the % symbol and multiplying. For example, $25\% = 25(0.01) = 0.25$. A shortcut for this process is as follows.

The decimal point in a whole number is to the immediate right of the rightmost digit.

Changing a Percent to Decimal Form

To change a percent to decimal form, move the decimal point two places to the left and drop the percent symbol.

Problem Change the percent to decimal form.

a. 75%

b. 35.5%

c. 6%

d. 300%

e. $33\frac{1}{3}\%$

Solution

a. 75%

Step 1. Move the decimal point two places to the left and drop the percent symbol.

$$75\% = 0.\underset{\leftarrow}{75} = 0.75$$

(two places left)

Step 2. State the main result.

$$75\% = 0.75$$

b. 35.5%

Step 1. Move the decimal point two places to the left and drop the percent symbol.

$$35.5\% = 0.\underset{\leftarrow}{355} = 0.355$$

(two places left)

Step 2. State the main result.

$$35.5\% = 0.355$$

c. 6%

Step 1. Move the decimal point two places to the left and drop the percent symbol.

$$6\% = 0.\underset{\leftarrow}{06} = 0.06$$

(two places left)

You may have to insert zeros in order to move two decimal places to the left.

Step 2. State the main result.

$$6\% = 0.06$$

6% ≠ 0.6. Be sure to move the decimal point *two* places to the left when changing from a percent to a decimal number.

d. 300%

Step 1. Move the decimal point two places to the left and drop the percent symbol.

$$300\% = 3.\underset{\leftarrow}{00} = 3$$

(two places left)

Step 2. State the main result.

$$300\% = 3$$

e. $33\frac{1}{3}\%$

Step 1. Move the decimal point two places to the left and drop the percent symbol.

$$33\frac{1}{3}\% = 0.\underset{\leftarrow}{33}\frac{1}{3} = 0.33\frac{1}{3}$$

(two places left)

A fraction in a decimal number does not occupy its own decimal place. It occupies the same decimal place as the digit to which it is attached.

Step 2. State the main result.

$$33\frac{1}{3}\% = 0.33\frac{1}{3}$$

Changing Decimals to Percent Form

You can reverse the process just presented to change decimals to percent form as follows.

Changing a Decimal to Percent Form

To change a decimal number to percent form, move the decimal point two places to the right and attach a % symbol.

The % symbol is equivalent to two decimal places, so when your number "gives up" two decimal places, you replace them with a % symbol.

Problem Change the decimal number to percent form.

a. 0.34

b. 0.425

c. 0.09

d. 2

e. 0.0025

Solution

a. 0.34

Step 1. Move the decimal point two places to the right and attach a % symbol.

$$0.34 = 0\underset{\rightarrow}{34}.\% = 34\%$$

(two places right)

Step 2. State the main result.

$$0.34 = 34\%$$

b. 0.425

Step 1. Move the decimal point two places to the right and attach a % symbol.

$$0.425 = 0\underset{\rightarrow}{42}.5\% = 42.5\%$$

(two places right)

Step 2. State the main result.

$$0.425 = 42.5\%$$

c. 0.09

Step 1. Move the decimal point two places to the right and attach a % symbol.

$$0.09 = 0\underset{\rightarrow}{09}.\% = 9\%$$

(two places right)

Step 2. State the main result.

$$0.09 = 9\%$$

d. 2

Step 1. Move the decimal point two places to the right and attach a % symbol.

$$2 = 2. = 2\underset{\rightarrow}{00}.\% = 200\%$$

(two places right)

You may have to insert zeros in order to move the decimal point two places to the right.

Step 2. State the main result.

$$2 = 200\%$$

e. 0.0025

Step 1. Move the decimal point two places to the right and attach a % symbol.

$$0.0025 = 0\underrightarrow{00}.25\% = 0.25\%$$
(two places right)

Step 2. State the main result.

$$0.0025 = 0.25\%$$

Changing Percents to Fraction Form

To change a percent to fraction form, you substitute $\frac{1}{100}$ for the % symbol and multiply.

Changing a Percent to Fraction Form

To change a percent to fraction form, substitute $\frac{1}{100}$ for the % symbol and then multiply. Simplify the resulting fraction to lowest terms, if possible.

Problem Change the percent to lowest fraction form.

a. 5%

b. 25%

c. 50%

d. 100%

e. 125%

f. $\frac{1}{4}\%$

Solution

a. 5%

Step 1. Substitute $\frac{1}{100}$ for the % symbol and then multiply.

$$5\% = 5 \cdot \frac{1}{100} = \frac{5}{100}$$

Step 2. Reduce to lowest form.

$$\frac{5}{100} = \frac{5 \div 5}{100 \div 5} = \frac{1}{20}$$

Step 3. State the main results.

$$5\% = \frac{5}{100} = \frac{1}{20}$$

b. 25%

Step 1. Substitute $\frac{1}{100}$ for the % symbol and multiply.

$$25\% = 25 \cdot \frac{1}{100} = \frac{25}{100}$$

Step 2. Reduce to lowest form.

$$\frac{25}{100} = \frac{25 \div 25}{100 \div 25} = \frac{1}{4}$$

Step 3. State the main results.

$$25\% = \frac{25}{100} = \frac{1}{4}$$

c. 50%

Step 1. Substitute $\frac{1}{100}$ for the % symbol and multiply.

$$50\% = 50 \cdot \frac{1}{100} = \frac{50}{100}$$

Step 2. Reduce to lowest form.

$$\frac{50}{100} = \frac{50 \div 50}{100 \div 50} = \frac{1}{2}$$

Step 3. State the main results.

$$50\% = \frac{50}{100} = \frac{1}{2}$$

d. 100%

Step 1. Substitute $\frac{1}{100}$ for the % symbol and multiply.

$$100\% = 100 \cdot \frac{1}{100} = \frac{100}{100}$$

Step 2. Reduce to lowest form.

$$\frac{100}{100} = 1$$

Step 3. State the main results.

$$100\% = \frac{100}{100} = 1$$

e. 125%

Step 1. Substitute $\frac{1}{100}$ for the % symbol and multiply.

$$125\% = 125 \cdot \frac{1}{100} = \frac{125}{100}$$

Step 2. Reduce to lowest form.

$$\frac{125}{100} = \frac{125 \div 25}{100 \div 25} = \frac{5}{4} = 1\frac{1}{4}$$

Step 3. State the main results.

$$125\% = \frac{125}{100} = \frac{5}{4} = 1\frac{1}{4}$$

f. $\frac{1}{4}\%$

Step 1. Substitute $\frac{1}{100}$ for the % symbol and multiply.

$$\frac{1}{4} \times \frac{1}{100} = \frac{1}{400}$$

Step 2. State the main results.

$$\frac{1}{4}\% = \frac{1}{400}$$

In changing a percent to a fraction, if the percent contains a mixed fraction, change the mixed fraction to an improper fraction as an initial step.

Problem Change the percent to lowest fraction form.

a. $12\frac{1}{2}\%$

b. $33\frac{1}{3}\%$

Solution

a. $12\frac{1}{2}\%$

Step 1. Change $12\frac{1}{2}$ to an improper fraction.

$$12\frac{1}{2}\% = \frac{25}{2}\%$$

Step 2. Substitute $\frac{1}{100}$ for the % symbol and multiply.

$$\frac{25}{2}\% = \frac{25}{2} \times \frac{1}{100} = \frac{25}{200}$$

Step 3. Reduce to lowest form.

$$\frac{25}{200} = \frac{25 \div 25}{200 \div 25} = \frac{1}{8}$$

Step 4. State the main results.

$$12\frac{1}{2}\% = \frac{25}{200} = \frac{1}{8}$$

b. $33\frac{1}{3}\%$

Step 1. Change $33\frac{1}{3}$ to an improper fraction.

$$33\frac{1}{3}\% = \frac{100}{3}\%$$

Step 2. Substitute $\frac{1}{100}$ for the % symbol and multiply.

$$\frac{100}{3}\% = \frac{100}{3} \times \frac{1}{100} = \frac{100}{300}$$

Step 3. Reduce to lowest form.

$$\frac{100}{300}=\frac{100\div100}{300\div100}=\frac{1}{3}$$

Step 4. State the main results.

$$33\frac{1}{3}\%=\frac{100}{300}=\frac{1}{3}$$

In changing a percent to a fraction, if the percent contains a decimal fraction, *before reducing to lowest form*, multiply the numerator and denominator by an appropriate power of 10 (10, 100, 1000, etc.) to remove the decimal in the numerator. Then simplify the resulting fraction, if possible.

Problem Change the percent to lowest fraction form.

a. 12.5%

b. 0.25%

Solution

a. 12.5%

Step 1. Substitute $\frac{1}{100}$ for the % symbol and multiply.

$$12.5\%=12.5\times\frac{1}{100}=\frac{12.5}{100}$$

Step 2. Multiply the numerator and denominator by 10 to remove the decimal in the numerator.

$$\frac{12.5}{100}=\frac{12.5\times10}{100\times10}=\frac{125}{1{,}000}$$

Step 3. Reduce to lowest form.

$$\frac{125}{1{,}000}=\frac{125\div125}{1{,}000\div125}=\frac{1}{8}$$

Step 4. State the main results.

$$12.5\%=\frac{125}{1{,}000}=\frac{1}{8}$$

b. 0.25%

Step 1. Substitute $\frac{1}{100}$ for the % symbol and multiply.

$$0.25\% = 0.25 \times \frac{1}{100} = \frac{0.25}{100}$$

Step 2. Multiply the numerator and denominator by 100 to remove the decimal in the numerator.

$$\frac{0.25}{100} = \frac{0.25 \times 100}{100 \times 100} = \frac{25}{10{,}000}$$

Step 3. Reduce to lowest form.

$$\frac{25}{10{,}000} = \frac{25 \div 25}{10{,}000 \div 25} = \frac{1}{400}$$

Step 4. State the main results.

$$0.25\% = \frac{25}{10{,}000} = \frac{1}{400}$$

Changing Fractions to Percent Form

An efficient way to change fractions to percent form is to first change the fractions to decimals. Here is the process.

Changing a Fraction to Percent Form

To change a fraction to percent form, first convert the fraction to a decimal number by performing the indicated division to *at least* two decimal places and then change the resulting decimal number to percent form. When the quotient is a repeating decimal number, write the remainder after two decimal places as a fraction like this: $\frac{\text{remainder}}{\text{divisor}}$.

Note: Even if the decimal representation terminates before two decimal places, you should carry the division to at least two decimal places.

Problem Change the fraction to percent form.

a. $\frac{4}{5}$

b. $\frac{3}{8}$

c. $\frac{1}{400}$

d. $\frac{1}{3}$

Solution

a. $\frac{4}{5}$

Step 1. Perform the indicated division.

$$\frac{4}{5} = 5\overline{)4.00} = 0.80 \quad (\text{quotient } 0.80)$$

Step 2. Change 0.80 to percent form.

$$0.80 = 0\underset{\rightarrow}{80}.\% = 80\%$$

(two places right)

Step 3. State the main results.

$$\frac{4}{5} = 0.80 = 80\%$$

b. $\frac{3}{8}$

Step 1. Perform the indicated division.

$$\frac{3}{8} = 8\overline{)3.000} = 0.375 \quad (\text{quotient } 0.375)$$

Step 2. Change 0.375 to percent form.

$$0.375 = 0\underset{\rightarrow}{37}.5\% = 37.5\%$$

(two places right)

Step 3. State the main results.

$$\frac{3}{8} = 0.375 = 37.5\%$$

c. $\dfrac{1}{400}$

Step 1. Perform the indicated division.

$$\frac{1}{400} = 400\overline{)1.0000}^{\;0.0025} = 0.0025$$

Step 2. Change 0.0025 to percent form.

$$0.0025 = 000.25\% = 0.25\%$$

(two places right)

Step 3. State the main results.

$$\frac{1}{400} = 0.0025 = 0.25\%$$

d. $\dfrac{1}{3}$

Step 1. Perform the indicated division.

$$\frac{1}{3} = 3\overline{)1.00}^{\;0.33 \text{ Remainder } 1} = 0.33\frac{1}{3}$$

Step 2. Change $0.33\frac{1}{3}$ to percent form.

$$0.33\frac{1}{3} = 033\frac{1}{3}.\% = 33\frac{1}{3}\%$$

(two places right)

Step 3. State the main results.

$$\frac{1}{3} = 0.33\frac{1}{3} = 33\frac{1}{3}\%$$

> $0.33\frac{1}{3} \neq 33.\frac{1}{3}\%$. The $\frac{1}{3}$ in $0.33\frac{1}{3}$ does not occupy a decimal place of its own.

Common Percents to Know

Here is a list of common percents and their decimal and fraction forms. Memorizing these relationships enhances your understanding of fractions, decimals, and percents.

$100\% = 1.00 = 1,\ \ 75\% = 0.75 = \frac{3}{4},\ \ 50\% = 0.50 = 0.5 = \frac{1}{2},\ \ 25\% = 0.25 = \frac{1}{4},$

$90\% = 0.90 = 0.9 = \frac{9}{10},\ \ 80\% = 0.80 = 0.8 = \frac{4}{5},\ \ 70\% = 0.70 = 0.7 = \frac{7}{10},$

$60\% = 0.60 = 0.6 = \frac{3}{5},\ \ 40\% = 0.40 = 0.4 = \frac{2}{5},\ \ 30\% = 0.30 = 0.3 = \frac{3}{10},$

$20\% = 0.20 = 0.2 = \frac{1}{5},\ \ 10\% = 0.10 = 0.1 = \frac{1}{10},\ \ 87\frac{1}{2}\% = 0.875 = \frac{7}{8},$

$62\frac{1}{2}\% = 0.625 = \frac{5}{8},\ \ 37\frac{1}{2}\% = 0.375 = \frac{3}{8},\ \ 12\frac{1}{2}\% = 0.125 = \frac{1}{8},$

$66\frac{2}{3}\% = 0.66\frac{2}{3} = \frac{2}{3},\ \ 33\frac{1}{3}\% = 0.33\frac{1}{3} = \frac{1}{3},$

$5\% = 0.05 = \frac{1}{20},\ \ 4\% = 0.04 = \frac{1}{25},\ \ 1\% = 0.01 = \frac{1}{100}$

Always remember that a % symbol means $\frac{1}{100}$ or 0.01.

Exercise 7

For 1–4, change the percent to decimal form.

1. 65%
2. 25.5%
3. 5%
4. 400%

For 5–9, change the decimal number to percent form.

5. 0.72
6. 0.325
7. 0.08
8. 10
9. 0.0075

For 10–13, change the percent to lowest fraction form.

10. 150%
11. $\frac{1}{2}\%$
12. 0.75%
13. $66\frac{2}{3}\%$

For 14 and 15, change the fraction to percent form.

14. $\frac{3}{5}$
15. $\frac{2}{3}$

8

Units of Measurement

In use in the United States are two different measurement systems: the US customary system and the metric system. In this chapter, you learn how to work with both systems.

Most other countries have adopted the metric system, although the imperial system—which is similar to the US customary system—is also used by the public in the United Kingdom.

Metric System Prefixes

You are no doubt familiar with the US customary system. Table 8.1 contains the metric system prefixes you will find useful to know.

Table 8.1 Metric System Prefixes

		MULTIPLES	
PREFIX	***ABBREVIATION***	***MEANING***	***POWER OF 10***
deka- or deca-	da	10 times	10^1
hecto-	h	100 times	10^2
kilo-	k	1000 times	10^3
mega-	M	1,000,000 times	10^6
		FRACTIONS	
PREFIX	***ABBREVIATION***	***MEANING***	***POWER OF 10***
deci-	d	0.1 times	10^{-1}
centi-	c	0.01 times	10^{-2}
milli-	m	0.001 times	10^{-3}
micro-	μ	0.000001 times	10^{-6}

US Customary and Metric Units

Table 8.2 contains a list of the US customary and metric units most commonly used in everyday activities.

Table 8.2 Units of Measurement

UNITS OF LENGTH

US SYSTEM	***METRIC SYSTEM***
1 foot (ft) = 12 inches (in)	1 centimeter (cm) = 10 millimeters (mm)
1 yard (yd) = 3 ft = 36 in	1 meter (m) = 100 centimeters
1 mile (mi) = 1,760 yd = 5,280 ft	1 kilometer (km) = 1000 m

When displaying numbers in the US system, whole numbers of five or more digits are shown in groupings of three, separated by commas. Four digits can be displayed with or without a comma.

UNITS OF VOLUME AND CAPACITY

US SYSTEM	***METRIC SYSTEM***
1 tablespoon (tbsp) = 3 teaspoons (tsp)	1 liter (L) = 1000 milliliters (mL) = 1000 cubic centimeters (cc)
1 fluid ounce (fl oz) = 2 tbsp	1 mL = 1 cc
1 cup (c) = 8 fl oz = 16 tbsp	
1 pint (pt) = 2 c = 16 fl oz	
1 quart (qt) = 2 pt = 32 fl oz	
1 gallon (gal) = 4 qt = 128 fl oz	
1 cubic foot (ft^3) = 1,728 cubic inches (in^3)	
1 cubic yard (yd^3) = 27 ft^3	

When displaying numbers in the metric system, no commas are used. Whole numbers of five or more digits are shown in groupings of three, separated by spaces. Four digits are displayed without a space.

UNITS OF WEIGHT

US SYSTEM	***METRIC SYSTEM***
1 pound (lb) = 16 ounces (oz)	1 gram (g) = 1000 milligrams (mg)
1 ton (T) = 2,000 lb	1 kilogram (kg) = 1000 grams

UNITS OF AREA

US SYSTEM	***METRIC SYSTEM***
1 square foot (ft^2) = 144 in^2	1 square meter (m^2) = 10 000 square centimeters (cm^2)
1 square yard (yd^2) = 9 (ft^2)	1 square kilometer (km^2) = 1 000 000 m^2
1 acre = 4,840 (yd^2)	
1 square mile (mi^2) = 640 acres	

(continued)

Table 8.2 Units of Measurement (*continued*)

UNITS OF REGULAR TIME	*CLOCK TIME*
1 minute (min) = 60 seconds (sec)	15 minutes past the hour = quarter past the hour
1 hour (hr) = 60 min = 3,600 sec	30 minutes past the hour = half past the hour
1 day (d) = 24 hr	45 minutes past the hour = three-quarters past the hour
1 week (wk) = 7 days (d)	a.m. denotes time before noon (morning)
1 month (mo) = 30 days (for ordinary accounting)	p.m. denotes time at noon or afternoon
1 year (yr) = 12 months = 52 weeks (wk) = 365 days (for 1 common year)	12:00 a.m. = midnight 12:00 p.m. = noon
1 decade = 10 yr	
1 century = 100 yr	

MONEY

1 dollar ($) = $1.00 = 100 cents (¢) = 20 nickels = 10 dimes = 4 quarters = 2 half-dollars
1 half-dollar = 50¢ = $0.50
1 quarter = 25¢ = $0.25
1 dime = 10¢ = $0.10
1 nickel = 5¢ = $0.05

US CUSTOMARY UNITS TO METRIC UNITS	*METRIC UNITS TO US CUSTOMARY UNITS*
1 in = 2.54 cm (exact)	1 cm ≈ 0.3937 in
1 yd ≈ 0.9144 m	1 m ≈ 1.094 yd = 39.3701 in
1 mi ≈ 1.6093 km	1 km ≈ 0.6213 mi
1 qt ≈ 0.946 L	1 L ≈ 1.057 qt
1 gal ≈ 3.785 L	1 L ≈ 0.246 gal
1 kg ≈ 2.205 lb	1 lb ≈ 0.4536 kg = 453.6 g

Denominate Numbers

You express measurements using denominate numbers. A *denominate number* is a number with units attached. Numbers without units attached are *abstract numbers.*

Problem Identify the denominate numbers in the following list:

1,000, 1200 m, \$400, $\frac{3}{4}$, $\frac{3}{4}$ yd, 180 d, 4,840 yd^2, 8 gal, 5 kg, 3.785

Solution

Step 1. Recalling that denominate numbers have units attached, identify the denominate numbers in the list.

1200 m, \$400, $\frac{3}{4}$ yd, 180 d, 4,840 yd^2, 8 gal, and 5 kg

Converting Units of Denominate Numbers

Convert measurement units of denominate numbers to different measurement units by using "conversion fractions." You make conversion fractions from conversion facts given in tables like Table 8.2. You have two conversion fractions for each conversion fact. For example, for 1 yd = 3 ft, you have $\frac{1\text{ yd}}{3\text{ ft}}$ and $\frac{3\text{ ft}}{1\text{ yd}}$. Each of these fractions is equivalent to the number 1 because the numerator and denominator are different names for the same length. Therefore, multiplying a quantity by either of these fractions, does not change the value of the quantity.

To change the units of a denominate number to different units, multiply by the conversion fraction whose *denominator is the same as the units of the quantity to be converted.* When you multiply, the units you started out with will divide ("cancel") out, and you will be left with the new units.

When you're converting from one measurement unit to another, if the original units don't cancel out when you multiply, then you picked the wrong conversion fraction. Go back and do the multiplication over again with the other conversion fraction.

You always should assess your result to see if it makes sense. Here is a helpful guideline.

When you convert from a larger unit to a smaller unit, it will take more of the smaller units to equal the same amount. When you convert from a smaller unit to a larger unit, it will take less of the larger units to equal the same amount.

Problem Change the given amount to the units indicated.

a. 5 yd = ________ ft

b. 360 min = ________ hr

c. 2 yd^2 = ________ ft^2

d. $\frac{3}{4}$ yd = ________ ft

e. 7 qt = _______ gal

f. 8 in = _______ cm

g. 1200 m = _______ km

h. 5 kg = _______ g

Solution

a. 5 yd = _______ ft

Step 1. Using Table 8.2, determine the conversion fractions.

The conversion fractions are $\frac{1\text{ yd}}{3\text{ ft}}$ and $\frac{3\text{ ft}}{1\text{ yd}}$.

Step 2. Write 5 yd as a fraction with denominator 1, select the conversion fraction that has "yd" in the denominator, and then multiply.

$$\frac{5\,\text{yd}}{1}\times\frac{3\,\text{ft}}{\text{yd}}=\frac{5\,\cancel{\text{yd}}}{1}\times\frac{3\,\text{ft}}{\cancel{\text{yd}}}=\frac{15\,\text{ft}}{1}=15\,\text{ft}$$

(The "yd" units cancel out, leaving "ft" as the units for the answer.)

Step 3. State the main result.

5 yd = __15__ ft

Step 4. Assess the result.

A yard is longer than a foot, so the number in front of "ft" should be greater than the number in front of "yd."

b. 360 min = _______ hr

Step 1. Using Table 8.2, determine the conversion fractions.

The conversion fractions are $\frac{1\text{ hr}}{60\text{ min}}$ and $\frac{60\text{ min}}{1\text{ hr}}$.

Step 2. Write 360 min as a fraction with denominator 1, select the conversion fraction that has "min" in the denominator, and then multiply.

$$\frac{360\,\text{min}}{1}\times\frac{1\,\text{hr}}{60\,\text{min}}=\frac{360\,\cancel{\text{min}}}{1}\times\frac{1\,\text{hr}}{60\,\cancel{\text{min}}}=\frac{360\,\text{hr}}{60}=6\,\text{hr}$$

(The "min" units cancel out, leaving "hr" as the units for the answer.)

Step 3. State the main result.

360 min = __6__ hr

Step 4. Assess the result.

A minute is shorter than an hour, so the number in front of "hr" should be less than the number in front of "min."

c. 2 yd^2 = ________ ft^2

Step 1. Using Table 8.2, determine the conversion fractions.

The conversion fractions are $\frac{1\text{ yd}^2}{9\text{ ft}^2}$ and $\frac{9\text{ ft}^2}{1\text{ yd}^2}$.

Step 2. Write 2 yd^2 as a fraction with denominator 1, select the conversion fraction that has "yd^2" in the denominator, and then multiply.

1 $yd^2 \neq 3$ ft^2. Do not make this common error. A square yard measures 1 yd by 1 yd. Thus, 1yd^2 = 1 yd × 1 yd = 3 ft × 3 ft = 9 ft^2.

$$\frac{2\text{ yd}^2}{1}\times\frac{9\text{ ft}^2}{1\text{ yd}^2}=\frac{2\text{ yd}^2}{1}\times\frac{9\text{ ft}^2}{1\text{ yd}^2}=\frac{18\text{ ft}^2}{1}=18\text{ ft}^2$$

(The "yd^2" units cancel out, leaving "ft^2" as the units for the answer.)

Step 3. State the main result.

2 yd^2 = ___18___ ft^2

Step 4. Assess the result.

A square yard is larger than a square foot, so the number in front of "ft^2" should be greater than the number in front of "yd^2."

d. $\frac{3}{4}$ yd = ________ ft

Step 1. Using Table 8.2, determine the conversion fractions.

The conversion fractions are $\frac{1\text{ yd}}{3\text{ ft}}$ and $\frac{3\text{ ft}}{1\text{ yd}}$.

Step 2. Write $\frac{3}{4}$ yd with "yd" as part of the numerator, select the conversion fraction that has "yd" in the denominator, and then multiply.

$\frac{9\text{ ft}}{4}=2\frac{1}{4}$ ft or 2.25 ft. When you have a choice of expressing an answer using fractions or decimals, you should use decimals because working with decimals is easier when you use a calculator.

$$\frac{3\text{ yd}}{4}\times\frac{3\text{ ft}}{1\text{ yd}}=\frac{3\text{ yd}}{4}\times\frac{3\text{ ft}}{1\text{ yd}}=\frac{9\text{ ft}}{4}=2.25\text{ ft}$$

(The "yd" units cancel out, leaving "ft" as the units for the answer.)

Step 3. State the main result.

$\frac{3}{4}$ yd = ___2.25___ ft

Step 4. Assess the result.

A yard is longer than a foot, so the number in front of "ft" should be greater than the number in front of "yd."

e. 7 qt = ________ gal

Step 1. Using Table 8.2, determine the conversion fractions.

The conversion fractions are $\frac{1\text{ gal}}{4\text{ qt}}$ and $\frac{4\text{ qt}}{1\text{ gal}}$.

Step 2. Write 7 qt as a fraction with denominator 1, select the conversion fraction that has "qt" in the denominator, and then multiply.

$$\frac{7\text{qt}}{1}\times\frac{1\text{gal}}{4\text{qt}}=\frac{7\,\cancel{\text{qt}}}{1}\times\frac{1\text{gal}}{4\,\cancel{\text{qt}}}=\frac{7\text{gal}}{4}=1.75\text{ gal}$$

(The "qt" units cancel out, leaving "gal" as the units for the answer.)

Step 3. State the main result.

7 qt = ___1.75___ gal

Step 4. Assess the result.

A quart is less than a gallon, so the number in front of "gal" should be less than the number in front of "qt."

f. 8 in = ________ cm

Step 1. Using Table 8.2, determine the conversion fractions.

The conversion fractions are $\frac{1\text{ in}}{2.54\text{ cm}}$ and $\frac{2.54\text{ cm}}{1\text{ in}}$.

Step 2. Write 8 in as a fraction with denominator 1, select the conversion fraction that has "in" in the denominator, and then multiply.

$$\frac{8\text{in}}{1}\times\frac{2.54\text{ cm}}{1\text{ in}}=\frac{8\,\cancel{\text{in}}}{1}\times\frac{2.54\text{ cm}}{1\,\cancel{\text{in}}}=\frac{20.32\text{ cm}}{1}=20.32\text{ cm}$$

(The "in" units cancel out, leaving "cm" as the units for the answer.)

Step 3. State the main result.

8 in = <u>20.32</u> cm

Step 4. Assess the result.

An inch is longer than a centimeter, so the number in front of "cm" should be greater than the number in front of "in."

g. 1200 m = ________ km

Step 1. Using Table 8.2, determine the conversion fractions.

The conversion fractions are $\frac{1\text{ km}}{1000\text{ m}}$ and $\frac{1000\text{ m}}{1\text{ km}}$.

Step 2. Write 1200 m as a fraction with denominator 1, select the conversion fraction that has "m" in the denominator, and then multiply.

$$\frac{1200\text{ m}}{1}\times\frac{1\text{ km}}{1000\text{ m}}=\frac{1200\ \cancel{\text{m}}}{1}\times\frac{1\text{ km}}{1000\ \cancel{\text{m}}}=\frac{1200\text{ km}}{1000}=1.2\text{ km}$$

(The "m" units cancel out, leaving "km" as the units for the answer.)

Step 3. State the main result.

1200 m = <u>1.2</u> km

Step 4. Assess the result.

A meter is shorter than a kilometer, so the number in front of "km" should be less than the number in front of "m."

h. 5 kg = __________ g

Step 1. Using Table 8.2, determine the conversion fractions.

The conversion fractions are $\frac{1\text{ kg}}{1000\text{ g}}$ and $\frac{1000\text{ g}}{1\text{ kg}}$.

Step 2. Write 5 kg as a fraction with denominator 1, select the conversion fraction that has "kg" in the denominator, and then multiply.

$$\frac{5\text{ kg}}{1}\times\frac{1000\text{ g}}{1\text{ kg}}=\frac{5\ \cancel{\text{kg}}}{1}\times\frac{1000\text{ g}}{1\ \cancel{\text{kg}}}=\frac{5000\text{ g}}{1}=5000\text{ g}$$

(The "kg" units cancel out, leaving "g" as the units for the answer.)

Step 3. State the main result.

5 kg = <u>5000</u> g

Step 4. Assess the result.

A kilogram is heavier than a gram, so the number in front of "g" should be greater than the number in front of "kg."

Shortcut for Converting Within the Metric System

If the base unit in the problem is the meter, liter, or gram, you have a shortcut to convert within the metric system. Use the mnemonic "**K**ing **H**enry **D**oesn't **U**sually **D**rink **C**hocolate **M**ilk," which helps you remember the following metric prefixes:

kilo-, **h**ecto-, **d**eca-, (base) **u**nit (no prefix), **d**eci-, **c**enti-, **m**illi-

The metric system is a decimal-based system, so the prefixes are based on powers of 10. Convert from one unit to another by either multiplying or dividing by a power of 10. If you move from left to right on the above list, then *multiply* by the power of 10 that corresponds to the number of times you moved. If you move from right to left, then *divide* by the power of 10 that corresponds to the number of times you moved. *Note:* When you use this shortcut, don't carry the units along when you do the calculations.

Problem Change the given amount to the units indicated.

a. 1200 m = ________ km

b. 5 kg = ________ g

c. 3.5 L = ________ mL

Solution

a. 1200 m = ________ km

Step 1. Using the list of metric prefixes, determine how many moves you make, and in what direction, to go from meters to kilometers.

kilo-, hecto-, deca-, meter, deci-, centi-, milli-

← three moves left

Step 2. Divide 1,200 by 10^3, the power of 10 that corresponds to the number of moves.

$1{,}200 \div 10^3 \text{(three moves left)} = 1{,}200 \div 1{,}000 = 1.2$

Step 3. State the main result.

1200 m = ___1.2___ km

Step 4. Assess the result.

A meter is shorter than a kilometer, so the number in front of "km" should be less than the number in front of "m."

b. 5 kg = ________ g

Step 1. Using the list of metric prefixes, determine how many moves you make, and in what direction, to go from kilograms to grams.

kilo-, hecto-, deca-, gram, deci-, centi-, milli-

three moves right →

Step 2. Multiply 5 by 10^3, the power of 10 that corresponds to the number of moves.

$5 \times 10^3 \text{(three moves right)} = 5 \times 1{,}000 = 5{,}000$

Step 3. State the main result.

5 kg = ___5000___ g

Step 4. Assess the result.

A kilogram is heavier than a gram, so the number in front of "g" should be greater than the number in front of "kg."

c. 3.5 L = __________ mL

Step 1. Using the list of metric prefixes, determine how many moves you make, and in what direction, to go from liters to milliliters.

kilo-, hecto-, deca-, liter, deci-, centi-, milli-

3 moves right →

Step 2. Multiply 3.5 by 10^3, the power of 10 that corresponds to the number of moves.

$3.5 \times 10^3 \text{(three moves right)} = 3.5 \times 1{,}000 = 3{,}500$

Step 3. State the main result.

3.5 L = ___3500___ mL

Step 4. Assess the result.

A liter is larger than a milliliter, so the number in front of "mL" should be greater than the number in front of "L."

Using a "Chain" of Conversion Fractions

For some conversions, you may need to use a "chain" of conversion fractions to obtain your desired units. Select the conversion facts that help you obtain your desired units and then multiply one after the other.

Problem Change the given amount to the units indicated.

a. 4 gal = ________ pt

b. 1 wk = ________ min

Solution

a. 4 gal = ________ pt

Step 1. Using Table 8.2, determine the conversion fractions.

Table 8.2 does not have a fact that shows the equivalency between gallons and pints. However, the table shows that 1 qt = 2 pt and 1 gal = 4 qt. These two facts yield four conversion fractions: $\frac{1 \text{ qt}}{2 \text{ pt}}$ and $\frac{2 \text{ pt}}{1 \text{ qt}}$ and $\frac{1 \text{ gal}}{4 \text{ qt}}$ and $\frac{4 \text{ qt}}{1 \text{ gal}}$.

Step 2. Start with $\frac{4 \text{ gal}}{1}$ and keep multiplying by conversion fractions until you obtain your desired units.

$$\frac{4 \text{ gal}}{1} \times \frac{4 \text{ qt}}{1 \text{ gal}} \times \frac{2 \text{ pt}}{1 \text{ qt}} = \frac{4 \cancel{\text{gal}}}{1} \times \frac{4 \cancel{\text{qt}}}{1 \cancel{\text{gal}}} \times \frac{2 \text{ pt}}{1 \cancel{\text{qt}}} = 32 \text{ pt}$$

Step 3. State the main result.

4 gal = ___32___ pt

Step 4. Assess the result.

A gallon is larger than a pint, so the number in front of "pt" should be greater than the number in front of "gal."

b. 1 wk = _________ min

Step 1. Using Table 8.2, determine the conversion fractions.

Table 8.2 does not have a fact that shows the equivalency between weeks and minutes. However, the table shows that 1 wk = 7 d, 1 d = 24 hr, and 1 hr = 60 min. These three facts yield six conversion fractions: $\frac{1\text{ wk}}{7\text{ d}}$ and $\frac{7\text{ d}}{1\text{ wk}}$, $\frac{1\text{ d}}{24\text{ hr}}$ and $\frac{24\text{ hr}}{1\text{ d}}$, and $\frac{1\text{ hr}}{60\text{ min}}$ and $\frac{60\text{ min}}{1\text{ hr}}$.

Step 2. Start with $\frac{1\text{ wk}}{1}$ and keep multiplying by conversion fractions until you obtain your desired units.

$$\frac{1\text{ wk}}{1}\times\frac{7\text{ d}}{1\text{ wk}}\times\frac{24\text{ hr}}{1\text{ d}}\times\frac{60\text{ min}}{1\text{ hr}}=\frac{1\ \cancel{\text{wk}}}{1}\times\frac{7\ \cancel{\text{d}}}{1\ \cancel{\text{wk}}}\times\frac{24\ \cancel{\text{hr}}}{1\ \cancel{\text{d}}}\times\frac{60\text{ min}}{1\ \cancel{\text{hr}}}$$
$$=10{,}080\text{ min}$$

Step 3. State the main result.

1 wk = <u>10,080</u> min

Step 4. Assess the result.

A week is longer than a minute, so the number in front of "min" should be greater than the number in front of "wk."

Converting Money to Different Denominations

It is customary to speak of "denominations" of money rather than "units" of money.

When converting with denominations of money, it is helpful to change the original amount to cents and then to the denomination to which you are converting.

Problem Change the given amount to the denomination indicated.

a. 15 quarters = ________ nickels

b. 75 dimes = ________ quarters

Solution

a. 15 quarters = __________ nickels

Step 1. Convert 15 quarters to cents.

$$\frac{15 \text{ quarters}}{1} \times \frac{25¢}{\text{quarter}} = \frac{15 \cancel{\text{ quarters}}}{1} \times \frac{25¢}{\cancel{\text{quarter}}} - 375¢$$

Step 2. Convert 375¢ to nickels.

$$\frac{375¢}{1} \times \frac{1 \text{ nickel}}{5¢} = \frac{375\cancel{¢}}{1} \times \frac{1 \text{ nickel}}{5\cancel{¢}} = \frac{375 \text{ nickels}}{5} = 75 \text{ nickels}$$

Step 3. State the main result.

15 quarters = ___75___ nickels

Step 4. Assess the result.

Quarters have more value than nickels, so the number in front of "nickels" should be greater than the number in front of "quarters."

b. 75 dimes = __________ quarters

Step 1. Convert 75 dimes to cents.

$$\frac{75 \text{ dimes}}{1} \times \frac{10¢}{\text{dime}} = \frac{75 \cancel{\text{ dimes}}}{1} \times \frac{10¢}{\cancel{\text{dime}}} = 750¢$$

Step 2. Convert 750¢ to quarters.

$$\frac{750¢}{1} \times \frac{1 \text{ quarter}}{25¢} = \frac{750\cancel{¢}}{1} \times \frac{1 \text{ quarter}}{25\cancel{¢}} = \frac{750 \text{ quarters}}{25} = 30 \text{ quarters}$$

Step 3. State the main result.

75 dimes= ___30___ quarters

Step 4. Assess the result.

Dimes have less value than quarters, so the number in front of "quarters" should be less than the number in front of "dimes."

Rough Equivalencies for the Metric System

If you are not very familiar with the metric system, here are some "rough" equivalencies of the more common units for your general knowledge.

Approximate Metric Equivalencies

meter	About 3 inches longer than a yard
centimeter	About the width of a large paper clip
millimeter	About the thickness of a dime
kilometer	About five city blocks or a little farther than half a mile
liter	A little more than a quart
milliliter	A little less than one-fourth of a teaspoon
gram	About the weight of a small paper clip
milligram	About the weight of a grain of salt
kilogram	The weight of a liter of water or a little more than 2 pounds

Reading Measuring Instruments

The scale (or gauge) of a measuring instrument has tick marks that divide the scale into equal intervals. Usually, not every tick mark is labeled with a value. To read a measuring instrument, determine what each interval between tick marks on the measuring instrument represents.

Problem On the thermometer shown, what is the Fahrenheit (°F) temperature to the nearest degree?

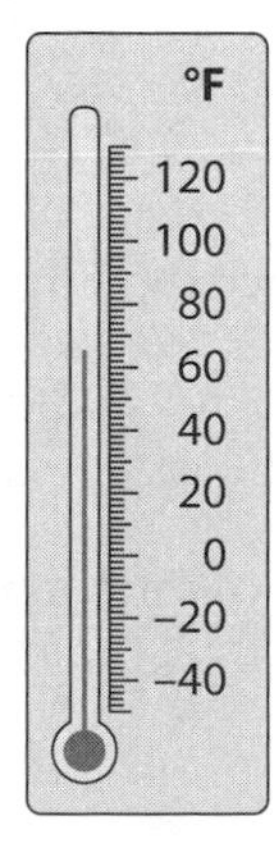

Step 1. Find the two consecutive labeled points immediately below and above the reading on the scale.

The thermometer is reading between 60° and 80°.

Step 2. Find the difference between the two consecutive labeled points.

The difference is 20° (= 80° – 60°).

Step 3. Count the number of tick marks to get from the lower point to the higher point. [Do not count the tick mark at the labeled lower point.]

There are 10 tick marks to get from 60° to 80°.

Step 4. Determine what each interval between tick marks on the thermometer represents.

Dividing 20°, the difference between the two points, by 10 yields 2° (= 20° ÷ 10). Therefore, each interval between tick marks on the thermometer represents 2°.

Step 5. Take the reading.

The thermometer is reading two tick marks above 60°. Since each interval between tick marks represents 2°, the thermometer is reading 4° above 60°, which is 64°F.

Determining Unit Price

Measurement skills include computing unit price. The *unit price* is the amount per unit. Unit price is used in many real-life situations.

Problem Which is a better buy for a certain product, 3 lb for \$2.00 or 4 lb for \$3.50?

Solution

Step 1. Compute the unit price for 3 lb for \$2.00.

The unit price for 3 lb for \$2.00 = $\frac{\$2.00}{3\text{ lb}}$ = \$0.67 per pound (rounded to the nearest cent).

Step 2. Compute the unit price for 4 lb for \$3.50.

The unit price for 4 lb for \$3.50 = $\frac{\$3.50}{4\text{ lb}}$ = \$0.88 per pound (rounded to the nearest cent).

Step 3. Compare the unit prices and select the better buy.

$0.67 per pound is less than $0.88 per pound, so, assuming the quality is the same, the better buy is 3 lb for $2.00.

Exercise 8

For 1–14, change the given amount to the units indicated.

1. 8.25 km = ________ m
2. 3 m = ________ cm
3. 3 hr = ________ min
4. 7,200 sec = ________ hr
5. 28 quarters = ________ dimes
6. 5 yd = ________ ft
7. 720 min = ________ hr
8. 3 yd^2 = ________ ft^2
9. 12 qt = ________ gal
10. 10 in = ________ cm
11. 7500 kg = ________ g
12. 8.5 L = ________ mL
13. 15 quarters = ________ nickels
14. $\frac{2}{3}$ yd = ________ ft
15. Which is a better buy for shelled peanuts, 8 oz for $4.50 or 9 oz for $4.99?
16. A runner ran 250 m. How many kilometers did the runner run?
17. First-class postage is charged by the ounce. A package weighs 3 lb 12 oz. How many ounces does the package weigh?
18. Carpet is sold by the square yard. The surface of the floor of a 9 ft by 12 ft room is 108 ft^2. How many square yards of carpet are needed to cover the floor?
19. A recipe calls for 3 tbsp of oil. How many fluid ounces is 3 tbsp?

20. A large container holds 5 gal of water. How many cups of water does the container hold?

21. On the thermometer shown, what is the Celsius (°C) temperature to the nearest degree?

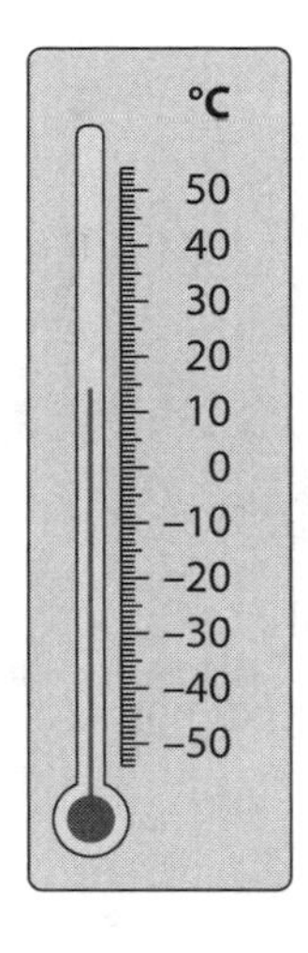

9

Ratios and Proportions

In this chapter, you learn about ratios and proportions.

Ratio Concepts

A *ratio* is a comparison by division of two quantities having the *same* units. Suppose the lengths of the radii of two circles are in the ratio of 3 to 5. If the radius of the first circle is 3 in, then the radius of the second circle is 5 in. That is, the radius of the first circle is $\frac{3}{5}$ as long as the radius of the second circle. Ratios can be expressed in several different forms: 3 to 5, 3 : 5, $3 \div 5$, $\frac{3}{5}$, 0.6, or 60%. The context of the discussion at the time usually dictates the form preferred. A ratio is a pure number—it has no units. The units "cancel" out as in $\frac{3\text{ in}}{5\text{ in}} = \frac{3}{5}$. Fractions, percents, and decimals are ratios. If the items compared have units that cannot be "canceled" such as $\frac{15\text{ mi}}{3\text{ hr}}$, then the comparison is called a *rate* or a *scale*.

Be careful when writing ratios. For instance, the comparison of 3 in to 5 ft is not a ratio because both are not expressed in the same units.

Proportion Concepts

The mathematical statement that two ratios (or rates or scales) are equal is a *proportion*. The statement $\frac{2}{3} = \frac{4}{6}$ is a proportion and is commonly read as "2 is to 3 as 4 is to 6." This can also be written as $2:3=4:6$, although for computations the fractional form is necessary. The fundamental property of proportions is that $\frac{a}{b}=\frac{c}{d}$ if and only if $ad=bc$. The products ad and bc are the *cross products*.

In a proportion, the cross products are equal to each other.

$$\frac{a}{b} \times \frac{c}{d}$$

Proportions and proportional thinking pervade such diverse topics as medicine dosage, application of fertilizer to a lawn, and oil additives in gasoline for outboard motors. Consequently, you should master the basics presented here to prepare for any real-world applications you might encounter.

Solving Proportions

If the values of three of the four terms of a proportion are known, then the value of the fourth term can be determined by using the fundamental property of proportions.

Problem Set up a proportion and solve for the unknown term.

a. A number *n* compared to 45 is the same as 7 compared to 15. Find *n*.

b. 38 is to 640 as 152 is to the number *n*. Find *n*.

Solution

a. A number *n* compared to 45 is the same as 7 compared to 15. Find *n*.

Step 1. Write a proportion.

$$\frac{n}{45}=\frac{7}{15}$$

Step 2. Solve the proportion.

Find a cross product you can calculate.

7×45

Divide by the numerical term you didn't use.

$$n=\frac{7\times 45}{15}$$

$n=21$

Step 3. Check.

$$\frac{n}{45} = \frac{7}{15}$$

$$\frac{21}{45} \stackrel{?}{=} \frac{7}{15}$$

$$\frac{7}{15} = \frac{7}{15} \quad \surd$$

b. 38 is to 640 as 152 is to the number *n*. Find *n*.

Step 1. Write a proportion.

$$\frac{38}{640} = \frac{152}{n}$$

Step 2. Solve the proportion.

Find a cross product you can calculate.

640×152

Divide by the numerical term you didn't use.

$$n = \frac{640 \times 152}{38}$$

$$n = 2{,}560$$

Step 3. Check.

$$\frac{38}{640} = \frac{152}{n}$$

$$\frac{38}{640} \stackrel{?}{=} \frac{152}{2{,}560}$$

$$\frac{38}{640} = \frac{38}{640} \quad \surd$$

Note: The fraction on the left initially could have been reduced.

It often simplifies the arithmetic in a problem to reduce as much as possible initially.

Solving Application Problems Involving Proportions

When you have an application problem involving proportions, look for a sentence or phrase in the problem that provides the information you need for the left portion of the proportion, and then look for another sentence or phrase that gives you the information you need for the right portion of the proportion.

Problem On a map, the distance between two cities is 13.5 in. On the map scale 0.5 in represents 20 mi. How far is it, in miles, between the two cities?

Solution This problem is a proportion problem involving a map scale. Information for the left portion of the proportion is in the first sentence of the problem, and information for the right portion of the proportion is in the second sentence.

Step 1. Use the first sentence to write the left portion of the proportion. Let d be the unknown distance. Then d corresponds to 13.5 in: $\frac{d}{13.5 \text{ in}}$.

Step 2. Use the second sentence to write the right portion of the proportion. 0.5 in corresponds to 20 mi: $\frac{20 \text{ mi}}{0.5 \text{ in}}$

Step 3. Write the proportion by setting the left portion equal to the right portion.

$$\frac{d}{13.5 \text{ in}} = \frac{20 \text{ mi}}{0.5 \text{ in}}$$

Notice that the units in the left portion match up with the units in the right portion.

Always check whether your units match up when you write a proportion. If you have miles in the numerator and inches in the denominator on the left, then you should have miles in the numerator and inches in the denominator on the right. If the units in the left and right portions don't match up, then your proportion is incorrect.

Step 4. Solve the proportion.

Find a cross product you can calculate.

$13.5 \text{ in} \times 20 \text{ mi}$

Divide by the numerical term you didn't use.

$$d = \frac{13.5 \text{ in} \times 20 \text{ mi}}{0.5 \text{ in}}$$

$$d = \frac{13.5 \cancel{\text{in}} \times 20 \text{ mi}}{0.5 \cancel{\text{in}}}$$

$d = 540 \text{ mi}$

Notice that the units work out to be miles, which is what you should expect because d is a distance.

In application problems, always check whether the units work out to be appropriate units for the unknown quantity.

Step 5. Check.

$$\frac{540 \text{ mi}}{13.5 \text{ in}} \stackrel{?}{=} \frac{20 \text{ mi}}{0.5 \text{ in}}$$

$$\frac{40 \text{ mi}}{1 \text{ in}} = \frac{40 \text{ mi}}{1 \text{ in}} \ \surd$$

You should always mentally check whether the answer makes sense. For instance, if 0.5 in represents 20 mi, then 1 in represents 40 mi, so 10 in should be 400 mi. Thus, an answer of 540 mi for 13.5 in does make sense.

Problem If 25 ft of wire weighs 3 lb, what is the weight of 8 ft of this wire?

Solution This problem is a proportion problem involving lengths and weights. Information for the left portion of the proportion is in the first part of the problem question, and information for the right portion of the proportion is in the second part of the problem question.

Step 1. Use the first part of the problem question to write the left portion of the proportion.

25 ft of wire corresponds to 3 lb: $\frac{25 \text{ ft}}{3 \text{ lb}}$

Step 2. Use the second part of the problem question to write the right portion of the proportion.

Let w be the unknown weight. Then 8 ft corresponds to w: $\frac{8 \text{ ft}}{w}$.

Step 3. Write the proportion by setting the left portion equal to the right portion.

$$\frac{25 \text{ ft}}{3 \text{ lb}} = \frac{8 \text{ ft}}{w}$$

Notice that the units in the left portion match up with the units in the right portion.

Step 4. Solve the proportion.

Find a cross product you can calculate.

(3 lb)(8 ft)

Divide by the numerical term you didn't use.

$$w = \frac{(3 \text{ lb})(8 \text{ ft})}{25 \text{ ft}}$$

$$w = \frac{(3 \text{lb})(8 \cancel{\text{ft}})}{25 \cancel{\text{ft}}}$$

$$w = 0.96 \text{ lb}$$

Notice that the units work out to be pounds, which is what you should expect because w is a weight.

Step 5. Check.

$$\frac{25 \text{ ft}}{3 \text{ lb}} \stackrel{?}{=} \frac{8 \text{ ft}}{0.96 \text{ lb}}$$

$$\frac{8.\overline{333} \text{ ft}}{1 \text{ lb}} = \frac{8.\overline{333} \text{ ft}}{1 \text{ lb}} \quad \checkmark$$

Problem The ratio of a propeller rate to that of the engine is 2 : 3. If the engine is turning at a rate of 4,800 revolutions per minute (rpm), what is the propeller rate?

Solution This is a proportion problem involving two ratios. The first ratio is in the first sentence, and the second ratio is in the question. Let R be the unknown propeller rate and compare the ratios.

Step 1. Write a proportion.

$$\frac{2}{3} = \frac{R}{4{,}800 \text{ rpm}}$$

Step 2. Solve the proportion

Find a cross product you can calculate.

$(2)(4{,}800 \text{ rpm})$

Divide by the numerical term you didn't use.

$$R = \frac{(2)(4{,}800 \text{ rpm})}{3}$$

$$R = 3{,}200 \text{ rpm}$$

Step 3. Check.

$$\frac{2}{3} \stackrel{?}{=} \frac{3{,}200 \text{ \cancel{rpm}}}{4{,}800 \text{ \cancel{rpm}}} = \frac{2}{3} \checkmark$$

Using Proportions to Solve Percent Problems

Percent problems can be solved using a "percent proportion," which has the following form:

$$\frac{\text{part}}{\text{whole}} = \frac{n}{100}$$

where

n = the number in front of the % sign
part = the quantity that is near the word *is* (when the word *is* occurs in the problem)
whole = the quantity that immediately follows the word *of*

The relationship among the three elements *n*, *part*, and *whole* can be explained in a percent statement like this:

The part is n% of the whole.

The secret to solving percent problems is being able to identify the three elements correctly. Start with n and the whole because they are usually easier to find. The part will be the other amount in the problem. The value of two of the elements will be given in the problem, and you will be solving for the third element. After you identify the three elements, substitute the two you know into the percent proportion and solve for the one that you don't know.

Problem Use a percent proportion to solve the given percent problem.

a. You inherited 15% of $9,000. How much money did you inherit?

b. A student scored 60 out of 80 questions. What is the student's percent grade?

c. A toy is on sale for $13. This sale price is 80% of the regular price of the toy. What is the regular price?

d. What is 30% of 140?

e. Twenty-five is 40% of what number?

f. Twenty-seven out of 40 is what percent?

Solution

a. You inherited 15% of $9,000. How much money did you inherit?

Step 1. Write the percent proportion.

$$\frac{\text{part}}{\text{whole}} = \frac{n}{100}$$

Step 2. Set up the percent proportion using $9,000 as the whole and 15 as n. Let p be the part inherited.

$$\frac{p}{\$9{,}000} = \frac{15}{100}$$

Step 3. Solve the proportion.

$$p = \frac{\$9,000 \times 15}{100} = \$1,350$$

Step 4. Check.

$$\frac{1,350}{9,000} \stackrel{?}{=} \frac{15}{100}$$

$0.15 = 0.15$ √

Step 5. Answer the question.

$1,350 is the amount inherited.

b. A student scored 60 out of 80 questions. What is the student's percent grade?

Step 1. Write the percent proportion.

$$\frac{\text{part}}{\text{whole}} = \frac{n}{100}$$

Step 2. Set up the percent proportion using 80 as the whole and 60 as the part. Let n be the percent number.

$$\frac{60}{80} = \frac{n}{100}$$

Step 3. Solve the proportion.

$$n = \frac{100(60)}{80} = 75$$

Thus, $n\% = 75\%$

Step 4. Check.

$$\frac{60}{80} \stackrel{?}{=} \frac{75}{100}$$

$0.75 = 0.75$ √

Step 5. Answer the question.

The percent score is 75%.

c. A toy is on sale for \$13. This sale price is 80% of the regular price of the toy. What is the regular price?

Step 1. Write the percent proportion.

$$\frac{\text{part}}{\text{whole}} = \frac{n}{100}$$

Step 2. Set up the percent proportion using the regular price as the whole and \$13 as the part. Let w be the regular price.

$$\frac{\$13}{w} = \frac{80}{100}$$

Step 3. Solve the proportion.

$$w = \frac{\$13(100)}{80} = \$16.25$$

Step 4. Check.

$$\frac{13}{16.25} \stackrel{?}{=} \frac{80}{100}$$

$$0.80 = 0.80 \checkmark$$

Step 5. Answer the question.

The regular price is \$16.25.

d. What is 30% of 140?

Step 1. Write the percent proportion.

$$\frac{\text{part}}{\text{whole}} = \frac{n}{100}$$

Step 2. Set up the percent proportion using 140 as the whole and $\frac{30}{100}$ as 30%. Let p be the part.

$$\frac{p}{140} = \frac{30}{100}$$

Step 3. Solve the proportion.

$$p = \frac{30(140)}{100} = 42$$

Step 4. Check.

$$\frac{42}{140} \stackrel{?}{=} \frac{30}{100}$$

$0.30 = 0.30$ √

Step 5. Answer the question.

30% of 140 is 42.

e. Twenty-five is 40% of what number?

Step 1. Write the percent proportion.

$$\frac{\text{part}}{\text{whole}} = \frac{n}{100}$$

Step 2. Set up the percent proportion using w as the whole, $\frac{40}{100}$ as 40%, and 25 as the part.

$$\frac{25}{w} = \frac{40}{100}$$

Step 3. Solve the proportion.

$$w = \frac{25(100)}{40} = 62.5$$

Step 4. Check.

$$\frac{25}{62.5} \stackrel{?}{=} \frac{40}{100}$$

$0.40 = 0.40$ √

Step 5. Answer the question.

25 is 40% of 62.5.

f. Twenty-seven out of 40 is what percent?

Step 1. Write the percent proportion.

$$\frac{\text{part}}{\text{whole}} = \frac{n}{100}$$

Step 2. Set up the percent proportion using 40 as the whole and 27 as the part. Let n be the percent number.

$$\frac{27}{40} = \frac{n}{100}$$

Step 3. Solve the proportion.

$$\frac{27}{40} = \frac{n}{100}$$

$$n = \frac{27(100)}{40} = 67.5$$

Thus, $n\% = 67.5\%$

Step 4. Check.

$$\frac{27}{40} \stackrel{?}{=} \frac{67.5}{100}$$

$$0.675 = 0.675 \ \surd$$

Step 5. Answer the question.

27 out of 40 is 67.5%.

Exercise 9

1. If 14 oz of salt is mixed with 5 oz of pepper, what is the ratio of salt to pepper?
2. Solve the proportion $\frac{n}{50} = \frac{70}{250}$ for n.
3. In a paint mixture that uses 2 parts of white paint to 5 parts of blue paint, how many quarts of white paint are needed to mix with 20 quarts of blue paint?
4. A stake 10 ft high casts a shadow 8 ft long at the same time that a tree casts a shadow 60 ft long. What is the height of the tree?
5. If 7 g of iron combines with 4 g of sulfur to form iron sulfide, how much sulfur will combine with 56 g of iron?
6. The tax on a property valued at $12,000 is $800. Assuming the tax rate is the same, what is the value of a property taxed at $1,100?
7. Ninety-two is 80% of what number?
8. What is 35% of 80?

9. You buy a $5,000 savings certificate that pays 4% simple annual interest. How much interest will you earn in 6 months?
10. Forty-five out of 120 is what percent?
11. If a jet plane can travel 4,500 km in 5 hr, how many kilometers can it travel in 25 min?
12. A map has a scale of 1 in = 15 mi. How much distance is represented by 18 in?
13. What percent of 7,580 is 454.8?
14. Calcium and chlorine combine in the weight ratio of 36:64. How much chlorine will combine with 5 g of calcium?

10

Roots and Radicals

In this chapter, you learn about roots and radicals.

Square Roots

You *square* a number by multiplying the number by itself. For instance, the square of 4 is $4 \times 4 = 16$. Also, the square of -4 is $-4 \times -4 = 16$. Thus, 16 is the result of squaring 4 or -4. The reverse of squaring is *finding the square root.* The two square roots of 16 are 4 and -4. Every positive number has two square roots that are equal in absolute value, but opposite in sign. The number 0 has only one square root, namely, 0.

The product of two negative numbers is *positive*.

When you are working with real numbers (which are the numbers you work with in this book), don't try to find square roots of negative numbers because not one real number will multiply by itself to give a negative number.

Problem Find the two square roots of the given number.

a. 25

b. 100

c. $\frac{4}{9}$

d. 0.49

Solution

a. 25

Step 1. Find a positive number whose square is 25.

$5 \times 5 = 25$, so 5 is the positive square root of 25.

Step 2. Find a negative number whose square is 25.

$-5 \times -5 = 25$, so -5 is the negative square root of 25.

Step 3. Write the two square roots of 25.

5 and -5 are the two square roots of 25.

b. 100

Step 1. Find a positive number whose square is 100.

$10 \times 10 = 100$, so 10 is the positive square root of 100.

Step 2. Find a negative number whose square is 100.

$-10 \times -10 = 100$, so -10 is the negative square root of 100.

Step 3. Write the two square roots of 100.

10 and -10 are the two square roots of 100.

c. $\frac{4}{9}$

Step 1. Find a positive number whose square is $\frac{4}{9}$.

$\frac{2}{3} \times \frac{2}{3} = \frac{4}{9}$, so $\frac{2}{3}$ is the positive square root of $\frac{4}{9}$.

Step 2. Find a negative number whose square is $\frac{4}{9}$.

$-\frac{2}{3} \times -\frac{2}{3} = \frac{4}{9}$, so $-\frac{2}{3}$ is the negative square root of $\frac{4}{9}$.

Step 3. Write the two square roots of $\frac{4}{9}$.

$\frac{2}{3}$ and $-\frac{2}{3}$ are the two square roots of $\frac{4}{9}$.

d. 0.49

Step 1. Find a positive number whose square is 0.49.

$(0.7)(0.7) = 0.49$, so 0.7 is the positive square root of 0.49.

Step 2. Find a negative number whose square is 0.49.

$(-0.7)(-0.7) = 0.49$, so -0.7 is the negative square root of 0.49.

Step 3. Write the two square roots of 0.49.

0.7 and -0.7 are the two square roots of 0.49.

Principal Square Roots and Radicals

You use the symbolism $\sqrt{16}$, read as "the square root of 16," to represent the positive square root of 16. Thus, $\sqrt{16} = 4$. This number is the *principal square root* of 16. Thus, the principal square root of 16 is 4. The symbol $\sqrt{\ }$ is the square root radical symbol. Using this notation, you indicate the negative square root of 16 as $-\sqrt{16}$. Thus, $-\sqrt{16} = -4$. The expression $\sqrt{16}$ is a *radical*. The number under the $\sqrt{\ }$ symbol is the *radicand*.

$\sqrt{-16} \neq -4$. $\sqrt{-16}$ is the square root of a negative number. No real number multiplies by itself to give -16.

As discussed earlier, every positive number has a positive and a negative square root. The positive square root is the *principal square root* of the number. The principal square root of 0 is 0. The $\sqrt{\ }$ symbol *always* designates the principal square root. Thus, $\sqrt{16} = 4$, not -4 or ± 4.

The principal square root is *always one* number and that number is either positive or 0.

The $\sqrt{\ }$ symbol *always* gives *one* number as the answer and that number is either positive or 0.

Problem Find the indicated root.

a. $\sqrt{81}$

b. $\sqrt{100}$

c. $\sqrt{\dfrac{4}{25}}$

d. $\sqrt{0.25}$

e. $\sqrt{0}$

f. $\sqrt{9+16}$

Solution

a. $\sqrt{81}$

Step 1. The principal square root of 81 is the positive square root of 81, so find the positive number whose square is 81.

$9 \times 9 = 81$, so 9 is the positive square root of 81.

Step 2. State the principal square root of 81.

$\sqrt{81} = 9$

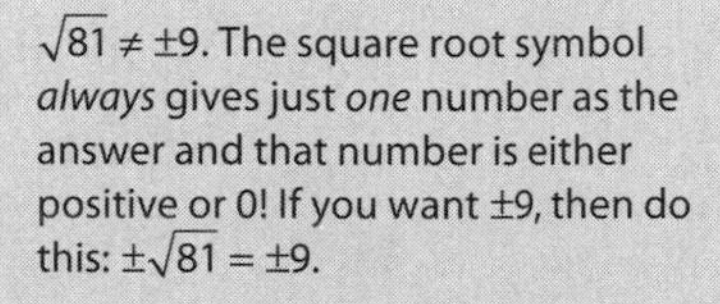

$\sqrt{81} \neq \pm 9$. The square root symbol *always* gives just *one* number as the answer and that number is either positive or 0! If you want ± 9, then do this: $\pm\sqrt{81} = \pm 9$.

b. $\sqrt{100}$

Step 1. The principal square root of 100 is the positive square root of 100, so find the positive number whose square is 100.

$10 \times 10 = 100$, so 10 is the positive square root of 10.

Step 2. State the principal square root of 100.

$\sqrt{100} = 10$

$\sqrt{100} \neq 50$. You do not divide by 2 to get a square root.

c. $\sqrt{\frac{4}{25}}$

Step 1. The principal square root of $\frac{4}{25}$ is the positive square root of $\frac{4}{25}$, so find the positive number whose square is $\frac{4}{25}$.

$\frac{2}{5} \times \frac{2}{5} = \frac{4}{25}$, so $\frac{2}{5}$ is the positive square root of $\frac{4}{25}$.

Step 2. State the principal square root of $\frac{4}{25}$.

$\sqrt{\frac{4}{25}} = \frac{2}{5}$

d. $\sqrt{0.25}$

Step 1. The principal square root of 0.25 is the positive square root of 0.25, so find the positive number whose square is 0.25.

$0.5 \times 0.5 = 0.25$, so 0.5 is the positive square root of 0.25.

Step 2. State the principal square root of 0.25.

$\sqrt{0.25} = 0.5$

e. $\sqrt{0}$

Step 1. State that the principal square root of 0 is 0.

$\sqrt{0} = 0$

f. $\sqrt{9+16}$

Step 1. Add 9 and 16 because you want the principal square root of the quantity $9+16$.

$\sqrt{9+16} = \sqrt{25}$

Always treat the $\sqrt{}$ symbol as a grouping symbol.

Step 2. The principal square root of 25 is the positive square root of 25, so find the positive number whose square is 25.

$5 \times 5 = 25$, so 5 is the positive square root of 25.

Step 3. State the principal square root.

$\sqrt{9+16} = \sqrt{25} = 5$

$\sqrt{9+16} \neq \sqrt{9} + \sqrt{16}$; $\sqrt{9+16} = \sqrt{25} = 5$, but $\sqrt{9} + \sqrt{16} = 3 + 4 = 7$.

Perfect Squares

A number that is an exact square of another number is a *perfect square*. For instance, 4, 9, 16, and 25 are perfect squares. Here is a helpful list of principal square roots of some perfect squares.

$\sqrt{0} = 0$, $\sqrt{1} = 1$, $\sqrt{4} = 2$, $\sqrt{9} = 3$, $\sqrt{16} = 4$, $\sqrt{25} = 5$, $\sqrt{36} = 6$,

$\sqrt{49} = 7$, $\sqrt{64} = 8$, $\sqrt{81} = 9$, $\sqrt{100} = 10$, $\sqrt{121} = 11$, $\sqrt{144} = 12$,

$\sqrt{169} = 13$, $\sqrt{196} = 14$, $\sqrt{225} = 15$,

$\sqrt{256} = 16$, $\sqrt{289} = 17$, $\sqrt{400} = 20$,

$\sqrt{625} = 25$

Working with square roots will be much easier for you if you memorize this list of square roots. Make flash cards to help you.

Also, fractions and decimals can be perfect squares. For instance, $\frac{9}{25}$ is a perfect square because $\frac{9}{25}$ equals $\frac{3}{5} \cdot \frac{3}{5}$, and 0.36 is a perfect square because 0.36 equals $(0.6)(0.6)$. If a number is not a perfect square, indicate its square roots by using the square root radical symbol. For instance, the two square roots of 15 are $\sqrt{15}$ and $-\sqrt{15}$.

Cube Roots

The product of a number used as a factor three times is the *cube* of that number. For instance, 64 is the cube of 4 because $4 \times 4 \times 4 = 64$ and, similarly, -64 is the cube of -4 because $-4 \cdot -4 \cdot -4 = -64$. The reverse of cubing is *finding the cube root*. Every number has one cube root, called its *principal cube root*. For example, because $4 \times 4 \times 4 = 64$, 4 is the principal cube root of 64. Likewise, because $-4 \cdot -4 \cdot -4 = -64$, -4 is the principal cube root of -64. As you can see, the principal cube root of a positive number is positive, and the principal cube root of a negative number is negative. Use the cube root radical symbol $\sqrt[3]{}$ (read as "the cube root of") to designate the principal cube root. The small number 3 in the symbol indicates that the cube root is desired. This number is the *index* of the radical. Thus, $\sqrt[3]{64} = 4$ and $\sqrt[3]{-64} = -4$.

Notice that you *can* find cube roots of negative numbers; negative numbers have negative cube roots.

If no index is written on a radical as in $\sqrt{}$, then the index is understood to be 2 and the radical indicates the principal square root.

Here is a list of principal cube roots of some *perfect cubes* that are useful to know.

You will find it worth your while to memorize this list of cube roots.

$$\sqrt[3]{0} = 0,\ \sqrt[3]{1} = 1,\ \sqrt[3]{8} = 2,\ \sqrt[3]{27} = 3,\ \sqrt[3]{64} = 4,$$

$$\sqrt[3]{125} = 5,\ \sqrt[3]{1{,}000} = 10$$

If a number is not a perfect cube, then indicate its principal cube root by using the cube root radical symbol. For instance, the principal cube root of -18 is $\sqrt[3]{-18}$.

Problem Find the indicated root.

a. $\sqrt[3]{27}$

b. $\sqrt[3]{-125}$

c. $\sqrt[3]{\frac{8}{125}}$

d. $\sqrt[3]{0.008}$

e. $\sqrt[3]{-1}$

Solution

a. $\sqrt[3]{27}$

Step 1. Find the positive number that you use as a factor three times to get 27.

$$3 \times 3 \times 3 = 27$$

Step 2. State the principal cube root of 27.

$$\sqrt[3]{27} = 3$$

b. $\sqrt[3]{-125}$

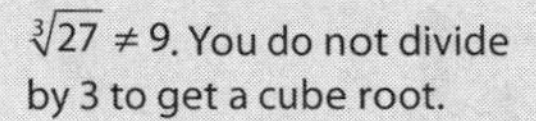

$\sqrt[3]{27} \neq 9$. You do not divide by 3 to get a cube root.

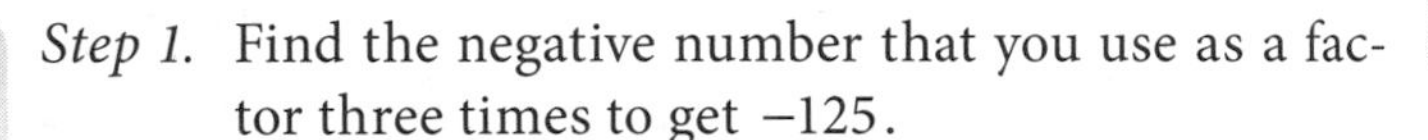

Step 1. Find the negative number that you use as a factor three times to get -125.

$$-5 \times -5 \times -5 = -125$$

Step 2. State the principal cube root of -125.

$$\sqrt[3]{-125} = -5$$

c. $\sqrt[3]{\frac{8}{125}}$

Step 1. Find the positive number that you use as a factor three times to get $\frac{8}{125}$.

$$\frac{2}{5} \times \frac{2}{5} \times \frac{2}{5} = \frac{8}{125}$$

Step 2. State the principal cube root of $\frac{8}{125}$.

$$\sqrt[3]{\frac{8}{125}} = \frac{2}{5}$$

d. $\sqrt[3]{0.008}$

Step 1. Find the positive number that you use as a factor three times to get 0.008.

$(0.2)(0.2)(0.2) = 0.008$

Step 2. State the principal cube root of 0.008.

$\sqrt[3]{0.008} = 0.2$

e. $\sqrt[3]{-1}$

Step 1. Find the negative number that you use as a factor three times to get -1.

$-1 \times -1 \times -1 = -1$

Step 2. State the principal cube root of -1.

$\sqrt[3]{-1} = -1$

Exercise 10

For 1–4, find the two square roots of the given number.

1. 144
2. $\frac{25}{49}$
3. 0.64
4. 400

For 5–10, find the indicated root, if possible.

5. $\sqrt{36}$
6. $\sqrt{-9}$
7. $\sqrt{\frac{16}{25}}$
8. $\sqrt{25+144}$
9. $\sqrt[3]{-8}$
10. $\sqrt[3]{\frac{64}{125}}$

11

Algebraic Expressions

This chapter presents a discussion of algebraic expressions. It begins with the basic terminology that is critical to your understanding of the concept of an algebraic expression.

Algebraic Terminology

A *variable* holds a place open for a number (or numbers, in some cases) whose value may vary. You usually express a variable as an upper- or lowercase letter (e.g., x, y, z, A, B, or C); for simplicity, the letter is the "name" of the variable. In problem situations, you use variables to represent unknown quantities. Although a variable may represent any number, in many problems, the variables represent specific numbers, but the values are unknown.

You can think of variables as numbers in disguise. Not recognizing that variables represent numbers is a common mistake.

A *constant* is a quantity that has a fixed, definite value that does not change in a problem situation. For example, all the real numbers are constants, including numbers whose units are units of measure such as 5 feet, 60 degrees, 100 pounds, and so forth. Also, the two special irrational numbers π and e are constants.

Recall that the real numbers are the natural numbers, the whole numbers, the integers, all positive and negative fractions and decimals, and all irrational numbers.

Even though the number π is represented by a Greek letter, π is not a variable. The number π is an irrational constant whose approximate value to two decimal places is 3.14. Similarly, the number e is an irrational constant whose approximate value to two decimal places is 2.72.

Problem Name the variable(s) and constant(s) in the given expression.

a. $\frac{5}{9}(F-32)$, where F is the number of degrees Fahrenheit

b. πd, where d is the measure of the diameter of a circle

Solution

a. $\frac{5}{9}(F-32)$, where F is the number of degrees Fahrenheit

Step 1. Recall that a letter names a variable whose value may vary.

Step 2. Name the variable(s).

The letter F stands for the number of degrees Fahrenheit and can be any number, and so it is a variable.

Step 3. Recall that a constant has a fixed, definite value.

Step 4. Name the constant(s).

The numbers $\frac{5}{9}$ and 32 have fixed, definite values that do not change, and so they are constants.

b. πd, where d is the measure of the diameter of a circle

Step 1. Recall that a letter names a variable whose value may vary.

Step 2. Name the variable(s).

The letter d stands for the measure of the diameter of a circle and can be any nonnegative number, and so it is a variable.

Step 3. Recall that a constant has a fixed, definite value.

Step 4. Name the constant(s).

The number π has a fixed, definite value that does not change, and so it is a constant.

If there is a number immediately next to a variable (normally, preceding it), that number is the *numerical coefficient* of the variable. If there is no number written immediately next to a variable, it is understood that the numerical coefficient is 1.

Problem State the numerical coefficient of the variable.

a. $-5x$

b. x

c. $20x$

Solution

a. $-5x$

Step 1. Identify the numerical coefficient by observing that the number -5 immediately precedes the variable x.

-5 is the numerical coefficient of x.

b. x

Step 1. Identify the numerical coefficient by observing that no number is written immediately next to the variable x, so the numerical coefficient is understood to be 1.

1 is the numerical coefficient of x.

> The numerical coefficient of x is not 0. The numerical coefficient of x is understood to be 1.

c. $20x$

Step 1. Identify the numerical coefficient by observing that the number 20 immediately precedes the variable x.

20 is the numerical coefficient of x.

Writing variables and coefficients or two or more variables (with or without constants) side by side with no multiplication symbol in between is a way to show multiplication. Thus, $-5x$ means -5 times x, and $2xyz$ means 2 times x times y times z. Also, a number or variable written immediately next to a grouping symbol indicates multiplication. For instance, $6(x+1)$ means 6 times the quantity $(x+1)$, $7\sqrt{x}$ means 7 times $\sqrt{x}$, and $-1|-8|$ means -1 times $|-8|$.

Evaluating Algebraic Expressions

An *algebraic expression* is a symbolic representation of a number. It can contain constants, variables, and computation symbols. Here are examples of algebraic expressions:

$$-5x\,,\ 2xyz,\ \frac{6(x+1)}{7\sqrt{x}+1},\ \frac{-1|y|+5(x-y)}{z+1},\ -8xy^3+\frac{5}{2x^2}-27,\ 8a^3+64b^3,$$

$$x^2-x-12,\ \frac{1}{3}x^2z^3,\text{ and }-2x^5+5x^4-3x^3-7x^2+x+4$$

You don't know what number an algebraic expression represents because algebraic expressions always contain variables. However, if you are given numerical values for the variables, you can evaluate the algebraic expression by substituting the given numerical value for each variable and then performing the indicated operations, being sure to *follow the order of operations* as you proceed. (See Chapter 4 for a discussion of order of operations.)

Problem Find the value of the algebraic expression when $x = 4$, $y = -8$, and $z = -5$.

a. $-5x$

b. $2xyz$

c. $\frac{6(x+1)}{7\sqrt{x}+1}$

d. $\frac{-1|y|+5(x-y)}{z+1}$

e. x^2-x-12

Solution

a. $-5x$

Step 1. Substitute 4 for x in the expression $-5x$.

$-5x$

$= -5(4)$

Step 2. Perform the indicated multiplication.

$= -20$

Step 3. State the main result.

$-5x = -20$ when $x = 4$.

b. $2xyz$

Step 1. Substitute 4 for x, -8 for y, and -5 for z in the expression $2xyz$.

$$2xyz$$
$$= 2(4)(-8)(-5)$$

When you substitute negative values into an algebraic expression, enclose them in parentheses to avoid careless errors.

Step 2. Perform the indicated multiplication.

$$= 320$$

Step 3. State the main result.

$2xyz = 320$ when $x = 4$, $y = -8$, and $z = -5$.

c. $\dfrac{6(x+1)}{7\sqrt{x}+1}$

Step 1. Substitute 4 for x in the expression $\dfrac{6(x+1)}{7\sqrt{x}+1}$.

$$\frac{6(x+1)}{7\sqrt{x}+1}$$
$$= \frac{6(4+1)}{7\sqrt{4}+1}$$

Step 2. Evaluate the resulting expression.

$$= \frac{6(5)}{7\cdot 2+1}$$
$$= \frac{30}{14+1}$$
$$= \frac{30}{15}$$
$$= 2$$

When you work with algebraic expressions, use a raised dot (·) or parentheses [( )] instead of the times symbol (x) to show multiplication between numerical quantities.

$7\sqrt{4}+1 \neq 7\sqrt{5}$. The square root applies only to the 4.

Step 3. State the main result.

$\dfrac{6(x+1)}{7\sqrt{x}+1} = 2$ when $x = 4$.

d. $\dfrac{-1|y|+5(x-y)}{z+1}$

Step 1. Substitute 4 for x, -8 for y, and -5 for z in the expression

$$\frac{-1|y|+5(x\quad y)}{z+1}.$$

$$\frac{-1|y|+5(x-y)}{z+1}$$

$$=\frac{-1|-8|+5(4-(-8))}{(-5)+1}$$

Step 2. Evaluate the resulting expression.

$$=\frac{-1|-8|+5(4-(-8))}{(-5)+1}$$

$$=\frac{-1(8)+5(4+8)}{-5+1}$$

$$=\frac{-8+5(12)}{-4}$$

$$=\frac{-8+60}{-4}$$

$$=\frac{52}{-4}$$

$$=-13$$

Step 3. State the main result.

$$\frac{-1|y|+5(x-y)}{z+1}=-13 \text{ when } x=4,\ y=-8, \text{ and } z=-5.$$

e. x^2-x-12

Step 1. Substitute 4 for x in the expression x^2-x-12.

$$x^2-x-12$$

$$=(4)^2-(4)-12$$

Step 2. Simplify the resulting expression.

$$= (4)^2 - (4) - 12$$
$$= 16 - 4 - 12$$
$$= 0$$

Step 3. State the main result.

$x^2 - x - 12 = 0$ when $x = 4$.

Problem Evaluate $-2x^5 + 5x^4 - 3x^3 - 7x^2 + x + 4$ when $x = -1$.

Solution

Step 1. Substitute $x = -1$ for x in the expression $-2x^5 + 5x^4 - 3x^3 - 7x^2 + x + 4$.

$$-2x^5 + 5x^4 - 3x^3 - 7x^2 + x + 4$$
$$= -2(-1)^5 + 5(-1)^4 - 3(-1)^3 - 7(-1)^2 + (-1) + 4$$

Step 2. Evaluate the resulting expression.

$$= -2(-1) + 5(1) - 3(-1) - 7(1) + (-1) + 4$$
$$= 2 + 5 + 3 - 7 - 1 + 4$$
$$= 6$$

Watch your signs! It's easy to make careless errors when you are evaluating negative numbers raised to powers.

Step 3. State the main result.

$-2x^5 + 5x^4 - 3x^3 - 7x^2 + x + 4 = 6$ when $x = -1$.

Dealing with Parentheses

Frequently, algebraic expressions are enclosed in parentheses. It is important that you deal with parentheses correctly.

If no symbol or if a + symbol immediately precedes parentheses that enclose an algebraic expression, remove the parentheses and rewrite the algebraic expression without changing any signs.

Problem Remove parentheses: $(-8x + 5)$.

Solution

Step 1. Remove the parentheses without changing any signs.

$$(-8x + 5) = -8x + 5$$

If a – symbol immediately precedes parentheses that enclose an algebraic expression, remove the parentheses and the – symbol and rewrite the algebraic expression but with all the signs changed.

Problem Remove parentheses: $-(-8x+5)$.

Solution

Step 1. Remove the parentheses and the – symbol and rewrite the expression, but change all the signs.

$$-(-8x+5)=8x-5$$

$-(-8x+5)\neq 8x+5$. Change *all* the signs, not just the first one. This mistake is very common.

Problem Remove parentheses.

a. $10+(3x^3-7x^2+2x)$

b. $10-(3x^3-7x^2+2x)$

Solution

a. $10+(3x^3-7x^2+2x)$

Step 1. Remove the parentheses and rewrite the algebraic expression in parentheses without changing any signs.

$$10+(3x^3-7x^2+2x)=10+3x^3-7x^2+2x$$

b. $10-(3x^3-7x^2+2x)$

Step 1. Remove the parentheses and the – symbol and rewrite the expression in parentheses, but change all the signs.

$$10-(3x^3-7x^2+2x) = 10-3x^3+7x^2-2x$$

If a number immediately precedes (or immediately follows) parentheses that enclose an algebraic expression, apply the distributive property to remove the parentheses.

Recall that the distributive property is $a(b+c)=a\cdot b+a\cdot c$ and $(b+c)a=b\cdot a+c\cdot a$.

Problem Remove parentheses.

a. $2(x+5)$

b. $-2(x+5)$

c. $2+4(x+y)$

Solution

a. $2(x+5)$

Step 1. Apply the distributive property.

$$2(x+5)$$
$$= 2 \cdot x + 2 \cdot 5$$
$$= 2x + 10$$

$2(x+5) \neq 2x+5$. You must multiply the 5 by 2 as well.

b. $-2(x+5)$

Step 1. Apply the distributive property.

$$-2(x+5)$$
$$= -2 \cdot x + -2 \cdot 5$$
$$= -2x + -10$$
$$= -2x - 10$$

It's always correct to change $+-$ to simply $-$.

c. $2+4(x+y)$

Step 1. Apply the distributive property.

$$2+4(x+y)$$
$$= 2 + 4 \cdot x + 4 \cdot y$$
$$= 2 + 4x + 4y$$

$2+4(x+y) \neq 6(x+y)$ because, in the order of operations, you multiply before you add. $2+4(x+y) = 2+4x+4y$, but $6(x+y) = 6 \cdot x + 6 \cdot y = 6x+6y$.

Exercise 11

1. Name the variable(s) and constant(s) in the expression $4s$, where s is the measure of the side of a square.

For 2–4, state the numerical coefficient of the variable.

2. $-12x$
3. z
4. $\frac{2}{3}x$

For 5–11, evaluate the algebraic expression when $x = 9$, $y = -2$, and $z = -3$.

5. $-5x$
6. $2xyz$
7. $\dfrac{6(x+1)}{5\sqrt{x}-10}$
8. $\dfrac{-2|y|+5(2x-y)}{-6z+y^3}$
9. $x^2 - 8x - 9$
10. $2y + x(y - z)$
11. $(y+z)^{-3}$

For 12–15, remove parentheses.

12. $5(x+6)$
13. $-3(x+4)$
14. $12+(x^2+y)$
15. $8-(2x-4y)$

12

Formulas

In this chapter, you evaluate formulas.

Definition of Formula

A *formula* is a rule that shows the mathematical relationship that connects two or more variables. Formulas are used extensively in mathematics, the natural and social sciences, engineering, and numerous other areas in the real world. You use your skills in evaluating algebraic expressions to evaluate formulas.

Evaluating Formulas

The symbols and/or letters in a formula are variables (except for special constants like π and e). If you are given numerical values for the variables, you evaluate the formula by substituting the given numerical value for each variable and then performing the indicated operations, being sure to *follow the order of operations* as you proceed.

If units are involved, carry the units term along in the computation, when it makes sense to do so. When you add or subtract units, the sum or difference has the same units. When you square a unit (i.e., $\text{unit} \times \text{unit}$), you get unit^2. When you cube a unit (i.e., $\text{unit} \times \text{unit} \times \text{unit}$), you get unit^3.

Problem Evaluate the formula to find the value of the indicated variable.

a. Find C when $F = 212$ using the formula $C = \frac{5}{9}(F - 32)$.

b. Find P when $s = 8$ m using the formula $P = 4s$.

c. Find positive value c when $a = 8$ and $b = 15$ using the formula $c^2 = a^2 + b^2$.

d. Find A when $l = 10$ ft and $w = 4$ ft using the formula $A = lw$.

e. Find I when $P = \$3,000$, $r = 4\%$ per year, and $t = 5$ yr using the formula $I = Prt$.

f. Find A when $r = 6$ yd using the formula $A = \pi r^2$. Use $\pi = 3.14$.

g. Find V when $r = 2.5$ cm and $h = 11.5$ cm using the formula $V = \pi r^2 h$. Use $\pi = 3.14$.

h. Find d when $r = 70$ mi per hour and $t = 5$ hr using the formula $d = rt$.

Solution

a. Find C when $F = 212$ using the formula $C = \frac{5}{9}(F - 32)$.

Step 1. Substitute 212 for F in the formula $C = \frac{5}{9}(F - 32)$.

$$C = \frac{5}{9}(F - 32)$$

$$C = \frac{5}{9}(212 - 32)$$

Step 2. Evaluate.

$$C = \frac{5}{9}(212 - 32)$$

$$C = \frac{5}{9}(180)$$

$$C = 100$$

b. Find P when $s = 8$ m using the formula $P = 4s$.

Step 1. Substitute 8 m for s in the formula $P = 4s$.

$$P = 4s$$

$$P = 4(8 \text{ m})$$

Step 2. Evaluate.

$$P = 4(8 \text{ m})$$
$$P = 32 \text{ m}$$

c. Find positive value c when $a = 8$ and $b = 15$ using the formula $c^2 = a^2 + b^2$.

Step 1. Substitute 8 for a and 15 for b in the formula $c^2 = a^2 + b^2$.

$$c^2 = a^2 + b^2$$
$$c^2 = (8)^2 + (15)^2$$

Step 2. Evaluate.

$$c^2 = (8)^2 + (15)^2$$
$$c^2 = 64 + 225$$
$$c^2 = 289$$

To obtain c, you must find the positive square root of 289. From the list of square roots in Chapter 10, you have $\sqrt{289} = 17$.

Thus, $c = 17$.

d. Find A when $l = 10$ ft and $w = 4$ ft using the formula $A = lw$.

Step 1. Substitute 10 ft for l and 4 ft for w in the formula $A = lw$.

$$A = lw$$
$$A = (10 \text{ ft})(4 \text{ ft})$$

Step 2. Evaluate.

$$A = (10 \text{ ft})(4 \text{ ft})$$
$$A = 40 \text{ ft}^2$$

e. Find I when $P = \$3,000$, $r = 4\%$ per year, and $t = 5$ yr using the formula $I = Prt$.

Step 1. Substitute \$3,000 for P, $\frac{4\%}{\text{yr}}$ for r, and 5 yr for t in the formula $I = Prt$.

$$I = Prt$$
$$I = (\$3,000)\left(\frac{4\%}{\text{yr}}\right)(5 \text{ yr})$$

Put quantities that follow the word *per* in the denominator of a fraction.

Step 2. Evaluate.

$$I = (\$3{,}000)\left(\frac{4\%}{\text{yr}}\right)(5\text{ yr})$$

$$I = (\$3{,}000)\left(\frac{0.04}{\text{yr}}\right)(5\text{ yr})$$

$$I = (\$3{,}000)\left(\frac{0.04}{\cancel{\text{yr}}}\right)(5\ \cancel{\text{yr}})$$

$$I = \$600$$

Notice that the "yr" units divide out, leaving "\$" as the units for the answer.

f. Find A when $r = 6$ yd using the formula $A = \pi r^2$. Use $\pi \approx 3.14$.

Step 1. Substitute 6 yd for r and 3.14 for π in the formula $A = \pi r^2$.

$$A = \pi r^2$$
$$A \approx (3.14)(6\text{ yd})^2$$

Step 2. Evaluate.

$$A = (3.14)(36\text{ yd}^2)$$
$$A = 113.04\text{ yd}^2$$

g. Find V when $r = 2.5$ cm and $h = 11.5$ cm using the formula $V = \pi r^2 h$. Use $\pi \approx 3.14$.

Step 1. Substitute 2.5 cm for r, 11.5 cm for h, and 3.14 for π in the formula $V = \pi r^2 h$.

$$V = \pi r^2 h$$
$$V \approx (3.14)(2.5\text{ cm})^2(11.5\text{ cm})$$

Step 2. Evaluate.

$$V = (3.14)(6.25\text{ cm}^2)(11.5\text{ cm})$$
$$V = 225.6875\text{ cm}^3$$

$(\text{cm}^2)(\text{cm}) = \text{cm} \times \text{cm} \times \text{cm} = \text{cm}^3$.

h. Find d when $r = 70$ mi per hour and $t = 5$ hr using the formula $d = rt$.

Step 1. Substitute $\frac{70 \text{ mi}}{\text{hr}}$ for r and 5 hr for t in the formula $d = rt$.

$$d = rt$$

$$d = \left(\frac{70 \text{ mi}}{\text{hr}}\right)(5 \text{ hr})$$

Step 2. Evaluate.

$$d = \left(\frac{70 \text{ mi}}{\text{hr}}\right)(5 \text{ hr})$$

$$d = \left(\frac{70 \text{ mi}}{\cancel{\text{hr}}}\right)(5 \cancel{\text{hr}})$$

$$d = 350 \text{ mi}$$

Notice that the "hr" units divide out, leaving "mi" as the units for the answer.

Exercise 12

Evaluate the formula to find the value of the indicated variable.

1. Find C when $F = 32$ using the formula $C = \frac{5}{9}(F - 32)$.
2. Find P when $s = 10$ m using the formula $P = 4s$.
3. Find positive value c when $a = 7$ and $b = 24$ using the formula $c^2 = a^2 + b^2$.
4. Find A when $l = 15$ ft and $w = 6$ ft using the formula $A = lw$.
5. Find I when $P = \$5{,}000$, $r = 3\%$ per year, and $t = 10$ yr using the formula $I = Prt$.
6. Find A when $r = 5$ yd using the formula $A = \pi r^2$. Use $\pi = 3.14$.
7. Find V when $r = 10$ ft and $h = 30$ ft using the formula $V = \pi r^2 h$. Use $\pi = 3.14$.
8. Find A when $b = 16$ and $h = 20$ using the formula $A = \frac{1}{2}bh$.
9. Find V when $r = 9$ cm and $h = 15$ cm using the formula $V = \frac{1}{3}\pi r^2 h$. Use $\pi = 3.14$.
10. Find F when $C = -15$ using the formula $F = \frac{9}{5}C + 32$.

13

Polynomials

In this chapter, you learn about polynomials. The chapter begins with a discussion of the elementary concepts that you need to know to ensure your success when working with polynomials and ends with addition and subtraction of polynomials.

Terms and Monomials

In an algebraic expression, *terms* are the parts of the expression that are connected to the other parts by indicated addition (plus sign) or indicated subtraction (minus sign). If the algebraic expression has no indicated addition or subtraction, then the algebraic expression itself is a term.

Problem Identify the terms in the given expression.

a. $-8x+\frac{5}{2x^2}-27$

b. $3x^5$

Solution

a. $-8x+\frac{5}{2x^2}-27$

Step 1. The expression contains indicated addition and subtraction, so identify the quantities between the plus and minus signs.

The terms are $-8x$, $\frac{5}{2x^2}$, and 27.

b. $3x^5$

Step 1. There is no indicated addition or subtraction, so the expression is a term.

The term is $3x^5$.

A *monomial* is a special type of term that, when simplified, is a constant or a product of one or more variables raised to *nonnegative* integer powers, with or without an explicit coefficient.

In monomials, no variable divisors, negative exponents, or variables as radicands of simplified radicals are allowed.

Problem Specify whether the term is a monomial. Explain your answer.

a. $-8x$

b. $\frac{5}{2x^2}$

c. 0

d. $3x^5$

e. $\sqrt{41}$

f. $4x^{-3}y^2$

g. $20\sqrt{x}$

Solution

a. $-8x$

Step 1. Check whether $-8x$ meets the criteria for a monomial.

$-8x$ is a term that is a variable raised to a positive integer power of 1 (understood), with an explicit coefficient of -8, so it is a monomial.

b. $\frac{5}{2x^2}$

Step 1. Check whether $\frac{5}{2x^2}$ meets the criteria for a monomial.

$\frac{5}{2x^2}$ is a term, but it contains division by a variable, so it is not a monomial.

c. 0

Step 1. Check whether 0 meets the criteria for a monomial.

0 is a constant, so it is a monomial.

d. $3x^5$

Step 1. Check whether $3x^5$ meets the criteria for a monomial.

$3x^5$ is a term that is a variable raised to a positive integer power of 5, with an explicit coefficient of 3, so it is a monomial.

e. $\sqrt{41}$

Step 1. Check whether $\sqrt{41}$ meets the criteria for a monomial.

$\sqrt{41}$ is a constant, so it is a monomial.

f. $4x^{-3}y^2$

Step 1. Check whether $4x^{-3}y^2$ meets the criteria for a monomial.

$4x^{-3}y^2$ contains a negative exponent, so it is not a monomial.

g. $20\sqrt{x}$

Step 1. Check whether $20\sqrt{x}$ meets the criteria for a monomial.

$20\sqrt{x}$ is a term, but it contains a variable as the radicand of a simplified radical, so it is not a monomial.

The *constants* in monomials can be divisors, have negative exponents, or be radicands in a radical. For instance, $\frac{5^{-2}x^2}{2\sqrt{3}}$ is a monomial.

Polynomials

A *polynomial* consists of a single monomial or two or more monomials connected by plus or minus signs. A polynomial that has exactly one term is a *monomial.* A polynomial that has exactly two terms is a *binomial.* A polynomial that has exactly three terms is a *trinomial.* A polynomial that has more than three terms is just a general polynomial.

Problem State the most specific name for the given polynomial.

a. $x^2 - 1$

b. $8a^3 + 64b^3$

c. $x^2 + 4x - 12$

d. $\frac{1}{3}x^2z^3$

e. $-2x^5 + 5x^4 - 3x^3 - 7x^2 + x + 4$

Solution

a. $x^2 - 1$

Step 1. Count the terms of the polynomial.

$x^2 - 1$ has exactly two terms.

Step 2. State the specific name.

$x^2 - 1$ is a binomial.

b. $8a^3 + 64b^3$

Step 1. Count the terms of the polynomial.

$8a^3 + 64b^3$ has exactly two terms.

Step 2. State the specific name.

$8a^3 + 64b^3$ is a binomial.

c. $x^2 + 4x - 12$

Step 1. Count the terms of the polynomial.

$x^2 + 4x - 12$ has exactly three terms.

Step 2. State the specific name.

$x^2 + 4x - 12$ is a trinomial.

d. $\frac{1}{3}x^2z^3$

Step 1. Count the terms of the polynomial.

$\frac{1}{3}x^2z^3$ has exactly one term

Step 2. State the specific name.

$\frac{1}{3}x^2z^3$ is a monomial.

e. $-2x^5 + 5x^4 - 3x^3 - 7x^2 + x + 4$

Step 1. Count the terms of the polynomial.

$-2x^5 + 5x^4 - 3x^3 - 7x^2 + x + 4$ has exactly six terms.

Step 2. State the specific name.

$-2x^5 + 5x^4 - 3x^3 - 7x^2 + x + 4$ is a polynomial.

Like Terms

Monomials that are constants or monomials that have exactly the same variable factors (i.e., the same letters with the same corresponding exponents) are *like terms*. Like terms are the same except, perhaps, for their coefficients. Terms that are not like terms are *unlike terms*.

Problem State whether the given monomials are like terms. Explain your answer.

a. $-10x$ and $25x$

b. $4x^2y^3$ and $-7x^3y^2$

c. 100 and 45

d. 25 and $25x$

Solution

a. $-10x$ and $25x$

Step 1. Check whether $-10x$ and $25x$ meet the criteria for like terms.

$-10x$ and $25x$ are like terms because they are exactly the same except for their numerical coefficients.

b. $4x^2y^3$ and $-7x^3y^2$

Step 1. Check whether $4x^2y^3$ and $-7x^3y^2$ meet the criteria for like terms.

$4x^2y^3$ and $-7x^3y^2$ are not like terms because the corresponding exponents on x and y are not the same.

c. 100 and 45

Step 1. Check whether 100 and 45 meet the criteria for like terms.

100 and 45 are like terms because they are both constants.

d. 25 and $25x$

Step 1. Check whether 25 and $25x$ meet the criteria for like terms.

25 and $25x$ are not like terms because they do not contain the same variable factors.

Adding and Subtracting Monomials

Because variables are standing in for numbers, you rely on the properties of numbers to justify operations with polynomials. (See Chapter 1 for a discussion of the properties of numbers.)

Addition and Subtraction of Monomials

1. To add monomials that are like terms, add their numerical coefficients and use the sum as the coefficient of their common variable component.
2. To subtract monomials that are like terms, subtract their numerical coefficients and use the difference as the coefficient of their common variable component.
3. To add or subtract unlike terms, indicate the addition or subtraction.

Problem Simplify.

a. $-10x + 25x$

b. $4x^2y^3 - 7x^3y^2$

c. $9x^2 + 3x^2 - 7x^2$

d. $25 + 25x$

e. $5x^2 - 7x^2$

Solution

a. $-10x + 25x$

Step 1. Check for like terms.

$-10x$ and $25x$ are like terms.

Step 2. Add the numerical coefficients.

$-10 + 25 = 15$

Step 3. Use the sum as the coefficient of x.

$-10x + 25x = 15x$

$-10x + 25x \neq 15x^2$. In addition and subtraction, the exponents on your variables do not change.

b. $4x^2y^3 - 7x^3y^2$

Step 1. Check for like terms.

$4x^2y^3$ and $7x^3y^2$ are not like terms, so leave the problem as indicated subtraction: $4x^2y^3 - 7x^3y^2$.

c. $9x^2 + 3x^2 - 7x^2$

Step 1. Check for like terms.

$9x^2$, $3x^2$, and $7x^2$ are like terms.

Step 2. Combine the numerical coefficients.

$9 + 3 - 7 = 5$

Step 3. Use the result as the coefficient of x^2.

$9x^2 + 3x^2 - 7x^2 = 5x^2$

d. $25 + 25x$

Step 1. Check for like terms.

25 and $25x$ are not like terms, so leave the problem as indicated addition: $25 + 25x$.

$25 + 25x \neq 50x$. These are not like terms, so you cannot combine them into one single term.

e. $5x^2 - 7x^2$

Step 1. Check for like terms.

$5x^2$ and $7x^2$ are like terms.

Step 2. Subtract the numerical coefficients.

$5 - 7 = -2$

Step 3. Use the result as the coefficient of x^2.

$5x^2 - 7x^2 = -2x^2$

Simplifying Polynomial Expressions

When you have an assortment of like terms in the same expression, systematically combine matching like terms in the expression. (For example, you might proceed from left to right.) You are *simplifying* the expression when you do this. To organize the process, use the properties of numbers to rearrange the expression so that matching like terms are together (later, you might choose to do this step mentally). If the expression includes unlike terms, just indicate the sums or differences of such terms. To avoid sign errors as you work, *keep a − symbol with the number that follows it.*

Problem Simplify.

a. $4x^3+5x^2-10x+25+2x^3-7x^2-5$

b. $5x - 20x + 30 - 10x + 100 + 3x - 25$

Solution

a. $4x^3+5x^2-10x+25+2x^3-7x^2-5$

Step 1. Check for like terms.

The like terms are $4x^3$ and $2x^3$, $5x^2$ and $7x^2$, and 25 and 5.

Step 2. Rearrange the expression so that like terms are together.

$4x^3+5x^2-10x+25+2x^3-7x^2-5$

$=4x^3+2x^3+5x^2-7x^2-10x+25-5$

Step 3. Systematically combine matching like terms and indicate addition or subtraction of unlike terms.

$= 6x^3 + -2x^2 - 10x + 20$

$= 6x^3 - 2x^2 -10x + 20$

Step 4. Review the main results.

$4x^3+5x^2-10x+25+2x^3-7x^2-5$

$=4x^3+2x^3+5x^2-7x^2-10x+25-5$

$= 6x^3 - 2x^2 - 10x + 20$

When you are simplifying, rearranging so that like terms are together can be done mentally. However, actually writing out this step helps you avoid careless errors.

Remember, when rearranging, to keep a − symbol with the number that follows it.

Because +− is equivalent to −, it is customary to change + − to simply − when you are simplifying expressions.

You should write polynomial answers in descending powers of a variable.

b. $5x - 20x + 30 - 10x + 100 + 3x - 25$

Step 1. Check for like terms.

The like terms are $5x, 20x, 10x,$ and $3x$ and $30, 100,$ and 25.

Step 2. Rearrange the expression so that like terms are together.

$5x - 20x + 30 - 10x + 100 + 3x - 25$

$= 5x - 20x - 10x + 3x + 30 + 100 - 25$

Step 3. Systematically combine matching like terms and indicate addition or subtraction of unlike terms.

$= -22x + 105$

Step 4. Review the main results.

$5x - 20x + 30 - 10x + 100 + 3x - 25$

$= 5x - 20x - 10x + 3x + 30 + 100 - 25$

$= -22x + 105$

Adding Polynomials

Addition of polynomials involves adding like terms.

Addition of Polynomials

To add two or more polynomials, add like monomial terms and simply indicate addition of unlike terms.

Problem Perform the indicated addition.

a. $(9x^2 - 6x + 2) + (-7x^2 - 5x + 3)$

b. $(4x^3 + 3x^2 - x + 8) + (8x^3 + 2x - 10)$

Solution

a. $(9x^2 - 6x + 2) + (-7x^2 - 5x + 3)$

Step 1. Remove parentheses.

$(9x^2 - 6x + 2) + (-7x^2 - 5x + 3)$

$= 9x^2 - 6x + 2 - 7x^2 - 5x + 3$

Step 2. Rearrange the terms so that like terms are together. (You may do this step mentally.)

$= 9x^2 - 7x^2 - 6x - 5x + 2 + 3$

Step 3. Systematically combine matching like terms and indicate addition or subtraction of unlike terms.

$= 2x^2 -11x + 5$

Step 4. Review the main results.

$(9x^2 - 6x + 2) + (-7x^2 - 5x + 3)$

$= 9x^2 - 6x + 2 - 7x^2 - 5x + 3$

$= 9x^2 - 7x^2 - 6x - 5x + 2 + 3$

$= 2x^2 - 11x + 5$

b. $(4x^3 + 3x^2 - x + 8) + (8x^3 + 2x - 10)$

Step 1. Remove parentheses.

$$(4x^3 + 3x^2 - x + 8) + (8x^3 + 2x - 10)$$
$$= 4x^3 + 3x^2 - x + 8 + 8x^3 + 2x - 10$$

Step 2. Rearrange the terms so that like terms are together. (You may do this step mentally.)

$$= 4x^3 + 8x^3 + 3x^2 - x + 2x + 8 - 10$$

Step 3. Systematically combine matching like terms and indicate addition or subtraction of unlike terms.

$$= 12x^3 + 3x^2 + x - 2$$

Step 4. Review the main results.

$$(4x^3 + 3x^2 - x + 8) + (8x^3 + 2x - 10)$$
$$= 4x^3 + 3x^2 - x + 8 + 8x^3 + 2x - 10$$
$$= 4x^3 + 8x^3 + 3x^2 - x + 2x + 8 - 10$$
$$= 12x^3 + 3x^2 + x - 2$$

Subtracting Polynomials

Subtraction of polynomials relies on your skills in adding polynomials.

Subtraction of Polynomials

To subtract two polynomials, add the opposite of the second polynomial.

You can accomplish subtraction of polynomials by enclosing both polynomials in parentheses and then placing a minus symbol between them. Of course, make sure that the minus symbol precedes the polynomial that is being subtracted.

Problem Perform the indicated subtraction.

a. $(9x^2 - 6x + 2) - (-7x^2 - 5x + 3)$

b. $(4x^3 + 3x^2 - x + 8) - (8x^3 + 2x - 10)$

Solution

a. $(9x^2 - 6x + 2) - (-7x^2 - 5x + 3)$

Step 1. Remove parentheses.

$(9x^2 - 6x + 2) - (-7x^2 - 5x + 3)$

$= 9x^2 - 6x + 2 + 7x^2 + 5x - 3$

Be careful with the signs! Sign errors are common mistakes in simplifying. Be sure to change the sign of *every* term in the second polynomial.

Step 2. Rearrange the terms so that like terms are together. (You may do this step mentally.)

$= 9x^2 + 7x^2 - 6x + 5x + 2 - 3$

Step 3. Systematically combine matching like terms and indicate addition or subtraction of unlike terms.

$= 16x^2 - x - 1$

Step 4. Review the main results.

$(9x^2 - 6x + 2) - (-7x^2 - 5x + 3)$

$= 9x^2 - 6x + 2 + 7x^2 + 5x - 3$

$= 9x^2 + 7x^2 - 6x + 5x + 2 - 3$

$= 16x^2 - x - 1$

b. $(4x^3 + 3x^2 - x + 8) - (8x^3 + 2x - 10)$

Step 1. Remove parentheses.

$(4x^3 + 3x^2 - x + 8) - (8x^3 + 2x - 10)$

$= 4x^3 + 3x^2 - x + 8 - 8x^3 - 2x + 10$

Step 2. Rearrange the terms so that like terms are together. (You may do this step mentally.)

$= 4x^3 - 8x^3 + 3x^2 - x - 2x + 8 + 10$

Step 3. Systematically combine matching like terms and indicate addition or subtraction of unlike terms.

$= -4x^3 + 3x^2 - 3x + 18$

Step 4. Review the main results.

$$(4x^3 + 3x^2 - x + 8) - (8x^3 + 2x - 10)$$
$$= 4x^3 + 3x^2 - x + 8 - 8x^3 - 2x + 10$$
$$= 4x^3 - 8x^3 + 3x^2 - x - 2x + 8 + 10$$
$$= -4x^3 + 3x^2 - 3x + 18$$

Exercise 13

For 1–5, state the most specific name for the given polynomial.

1. $x^2 - x + 1$
2. $125x^3 - 64y^3$
3. $2x^2 + 7x - 4$
4. $-\frac{1}{3}x^5y^2$
5. $2x^4 + 3x^3 - 7x^2 - x + 8$

For 6–15, simplify.

6. $-15x + 17x$
7. $14xy^3 - 7x^3y^2$
8. $10x^2 - 2x^2 - 20x^2$
9. $10 + 10x$
10. $2 + 4(x + 5)$
11. $12x^3 - 5x^2 + 10x - 60 + 3x^3 - 7x^2 - 1$
12. $(10x^2 - 5x + 3) + (6x^2 + 5x - 13)$
13. $(20x^3 - 3x^2 - 2x + 5) + (9x^3 + x^2 + 2x - 15)$
14. $(10x^2 - 5x + 3) - (6x^2 + 5x - 13)$
15. $(20x^3 - 3x^2 - 2x + 5) - (9x^3 + x^2 + 2x - 15)$

14

Solving Equations

In this chapter, you learn about equations and how to solve them.

Equation Terminology

An *equation* is a statement of equality between two mathematical expressions. Equations may be true or false. For instance, the equation $9 + 7 = 16$ is true, but the equation $9 + 7 = 20$ is false. An equation has two sides. Whatever is on the left side of the equal sign is the *left side* (LS) of the equation, and whatever is on the right side of the equal sign is the *right side* (RS) of the equation.

A variable (or variables) might hold the place for numbers in an equation. For example, the equation

$$5x + 9 = 3x - 1$$

is an equation that has one variable, namely, x. Of course, you can have other letters (e.g., y, z, and A) that represent the variable in an equation.

Solving Equations

To solve an equation, proceed systematically to undo what has been done to the variable. In this discussion, for convenience, the variable is x.

The goal in solving an equation for the variable x is to get x by itself on only one side of the equation and with a coefficient of 1 (usually understood).

Getting x by itself on only one side of the equation and everything else on the other side is known as "isolating the variable."

You solve an equation using the properties of real numbers and simple algebraic tools. An equation is like a balance scale. To keep the equation in balance, when you do something to one side of the equation, you must do the same thing to the other side of the equation.

Tools for Solving Equations

- Add the same number to both sides.
- Subtract the same number from both sides.
- Multiply both sides by the same *nonzero* number.
- Divide both sides by the same *nonzero* number.

When you are solving an equation, it is important to remember that you must *never* multiply or divide both sides by 0!

You focus on the variable x while you work because what has been done to x determines the operation you choose to do. As you proceed step-by-step, you exploit the fact that addition and subtraction undo each other, and, similarly, multiplication and division undo one another. Here is a systematic approach.

Solving an Equation for the Variable *x*

1a. If fractions are involved, enclose numerators of more than one term in parentheses and then multiply each term on both sides of the equation by the least common multiple (LCM) of all the denominators.
1b. If the equation contains parentheses, remove them using the distributive property.
2. Combine like terms, if any, on each side of the equation.
3. If an x term appears on both sides of the equation, add or subtract the x term so that, after you simplify, the x appears on only one side of the equation.
4. Undo addition or subtraction and then simplify. If a number is added to the x term, subtract that number from both sides of the equation. If a number is subtracted from the x term, add that number to both sides of the equation.
5. Divide both sides of the equation by the coefficient of x.
6. Check your work.
7. Check your answer.

Note: Steps 1a and 1b are labeled this way because these two steps are interchangeable, depending on your judgment of which needs to be done first.

Problem Solve the equation for x.

a. $5x + 9 = 3x - 1$

b. $4(x - 6) = 40$

c. $-3x - 7 = 14$

d. $3x - 2 = 7 - 2x$

e. $\frac{x-3}{2} = \frac{2x+4}{5}$

f. $3\left(\frac{1}{3}x - 2\right) = 2\left(7 - \frac{1}{2}x\right)$

Solution

a. $5x + 9 = 3x - 1$

Step 1. An x term appears on both sides of the equation, so subtract $3x$ from the RS to remove it from that side. To maintain balance, subtract $3x$ from the LS, too.

$5x + 9 - \mathbf{3x} = 3x - 1 - \mathbf{3x}$

Step 2. Simplify both sides by combining like variable terms.

$\mathbf{2x} + 9 = -1$

Step 3. 9 is added to the x term, so subtract 9 from both sides.

$2x + 9 - \mathbf{9} = -1 - \mathbf{9}$

Step 4. Simplify both sides by combining constant terms.

$2x = \mathbf{-10}$

Step 5. You want the coefficient of x to be 1, so divide both sides by 2.

$\frac{2x}{\mathbf{2}} = \frac{-10}{\mathbf{2}}$

Step 6. Simplify.

$\mathbf{x = -5}$

Step 7. Check your work by reviewing steps 1–6.

$5x + 9 = 3x - 1$

$5x + 9 - 3x = 3x - 1 - 3x$

$2x + 9 = -1$

$2x + 9 - 9 = -1 - 9$

$2x = -10$

$\frac{2x}{2} = \frac{-10}{2}$

$x = -5$

Step 8. Check your answer by substituting −5 for x in the original equation, $5x + 9 = 3x - 1$.

Substitute −5 for x on the LS of the equation. $5x + 9 = 5(-5) + 9 = -25 + 9 = -16$. Similarly, on the RS, you have $3x - 1 = 3(-5) - 1 = -15 - 1 = -16$. Both sides equal −16, so −5 is the correct answer.

b. $4(x - 6) = 40$

Step 1. Use the distributive property to remove parentheses.

$\mathbf{4} \cdot \boldsymbol{x} - \mathbf{4} \cdot \mathbf{6} = 40$

$4x - 24 = 40$

> $4(x - 6) \neq 4x - 6$. Multiply both terms by 4.

Step 2. 24 is subtracted from the x term, so add 24 to both sides.

$4x - 24 + \mathbf{24} = 40 + \mathbf{24}$

Step 3. Simplify both sides by combining constant terms.

$4x = \mathbf{64}$

Step 4. You want the coefficient of x to be 1, so divide both sides by 4.

$\frac{4x}{\mathbf{4}} = \frac{64}{\mathbf{4}}$

Step 5. Simplify.

$\boldsymbol{x} = \mathbf{16}$

Step 6. Check your work by reviewing steps 1–5.

$4(x - 6) = 40$

$4 \cdot x - 4 \cdot 6 = 40$

$4x - 24 = 40$

$4x - 24 + 24 = 40 + 24$

$4x = 64$

$$\frac{4x}{4} = \frac{64}{4}$$

$x = 16$

Step 7. Check your answer by substituting 16 for x in the original equation, $4(x - 6) = 40$.

Substitute 16 for x on the LS of the equation: $4(x - 6) = 4(16 - 6) = 4(10) = 40$. On the RS, you have 40 as well. Both sides equal 40, so 16 is the correct answer.

c. $-3x - 7 = 14$

Step 1. 7 is subtracted from the x term, so add 7 to both sides.

$-3x - 7 + \mathbf{7} = 14 + \mathbf{7}$

Step 2. Simplify both sides by combining constant terms.

$-3x = \mathbf{21}$

Step 3. You want the coefficient of x to be 1, so divide both sides by -3.

$$\frac{-3x}{\mathbf{-3}} = \frac{21}{\mathbf{-3}}$$

Step 4. Simplify.

$\mathbf{x = -7}$

Step 5. Check your work by reviewing steps 1–4.

$-3x - 7 = 14$

$-3x - 7 + 7 = 14 + 7$

$-3x = 21$

$$\frac{-3x}{-3} = \frac{21}{-3}$$

$x = -7$

Step 6. Check your answer by substituting -7 for x in the original equation, $-3x - 7 = 14$.

Substitute -7 for x on the LS of the equation: $-3x - 7 = -3(-7) - 7 = 21 - 7 = 14$. On the RS, you have 14 as well. Both sides equal 14, so -7 is the correct answer.

d. $3x - 2 = 7 - 2x$

Step 1. An x term appears on both sides of the equation, so add $2x$ to the RS to remove it from that side. To maintain balance, add $2x$ to the LS, too.

$3x - 2 + \mathbf{2x} = 7 - 2x + \mathbf{2x}$

Step 2. Simplify both sides by combining like variable terms.

$\mathbf{5x} - 2 = 7$

Step 3. 2 is subtracted from the x term, so add 2 to both sides.

$5x - 2 + \mathbf{2} = 7 + \mathbf{2}$

Step 4. Simplify both sides by combining constant terms.

$5x = \mathbf{9}$

Step 5. You want the coefficient of x to be 1, so divide both sides by 5.

$$\frac{5x}{5} = \frac{9}{5}$$

Step 6. Simplify.

$\mathbf{x = 1.8}$

Step 7. Check your work by reviewing steps 1–6.

$3x - 2 = 7 - 2x$

$3x - 2 + 2x = 7 - 2x + 2x$

$5x - 2 = 7$

$5x - 2 + 2 = 7 + 2$

$5x = 9$

$$\frac{5x}{5} = \frac{9}{5}$$

$x = 1.8$

Step 8. Check your answer by substituting 1.8 for x in the original equation, $3x - 2 = 7 - 2x$.

Substitute 1.8 for x on the LS of the equation: $3x - 2 = 3(1.8) - 2 = 5.4 - 2 = 3.4$. Similarly, on the RS, you have $7 - 2x = 7 - 2(1.8) = 7 - 3.6 = 3.4$. Both sides equal 3.4, so 1.8 is the correct answer.

e. $\frac{x-3}{2} = \frac{2x+4}{5}$

Step 1. Eliminate fractions by multiplying both sides by 10, the least common multiple of 2 and 5. Write 10 as $\frac{10}{1}$ to avoid errors.

$$\frac{\mathbf{10}}{\mathbf{1}} \cdot \frac{(x-3)}{2} = \frac{\mathbf{10}}{\mathbf{1}} \cdot \frac{(2x+4)}{5}$$

Step 2. Simplify

$$\frac{{}^{5}\cancel{\mathbf{10}}}{1} \cdot \frac{(x-3)}{\cancel{2}_{1}} = \frac{{}^{2}\cancel{\mathbf{10}}}{1} \cdot \frac{(2x+4)}{\cancel{5}_{1}}$$

$\mathbf{5(x-3) = 2(2x+4)}$

$5x - 15 = 4x + 8$

Step 3. An x term appears on both sides of the equation, so subtract $4x$ from the RS to remove it from that side. To maintain balance, subtract $4x$ from the LS, too.

$5x - 15 - \mathbf{4x} = 4x + 8 - \mathbf{4x}$

Step 4. Simplify both sides by combining variable terms.

$x - 15 = 8$

Step 5. 15 is subtracted from the x term, so add 15 to both sides.

$x - 15 + \mathbf{15} = 8 + \mathbf{15}$

Step 6. Simplify both sides by combining constant terms.

$x = \mathbf{23}$

Step 7. Check your work by reviewing steps 1–6.

$$\frac{x-3}{2} = \frac{2x+4}{5}$$

$$\frac{10}{1} \cdot \frac{(x-3)}{2} = \frac{10}{1} \cdot \frac{(2x+4)}{5}$$

$$\frac{{}^{5}\cancel{10}}{1} \cdot \frac{(x-3)}{\cancel{2}_{1}} = \frac{{}^{2}\cancel{10}}{1} \cdot \frac{(2x+4)}{\cancel{5}_{1}}$$

$5(x-3) = 2(2x+4)$

$5x - 15 = 4x + 8$

$5x - 15 - 4x = 4x + 8 - 4x$

$x - 15 = 8$

$x - 15 + 15 = 8 + 15$

$x = 23$

Step 8. Check your answer by using 23 for x in the original equation, $\frac{x-3}{2} = \frac{2x+4}{5}$.

Put in 23 for x on the LS of the equation: $\frac{x-3}{2} = \frac{23-3}{2} = \frac{20}{2} = 10.$

Similarly, on the RS, you have $\frac{2x+4}{5} = \frac{2(23+4)}{5} = \frac{46+4}{5} = \frac{50}{5} = 10.$

Both sides equal 10, so 23 is the correct answer.

f. $3\left(\frac{1}{3}x - 2\right) = 2\left(7 - \frac{1}{2}x\right)$

Step 1. Use the distributive property to remove parentheses.

$\mathbf{3 \cdot \frac{1}{3}x - 3 \cdot 2 = 2 \cdot 7 - 2 \cdot \frac{1}{2}x}$

$x - 6 = 14 - x$

Step 2. An x term appears on both sides of the equation, so add x to the RS to remove it from that side. To maintain balance, add x to the LS, too.

$x - 6 + \mathbf{x} = 14 - x + \mathbf{x}$

Step 3. Simplify both sides by combining like variable terms.

$\mathbf{2x} - 6 = 14$

Step 4. 6 is subtracted from the x term, so add 6 to both sides.

$2x - 6 + \mathbf{6} = 14 + \mathbf{6}$

Step 5. Simplify both sides by combining constant terms.

$2x = \mathbf{20}$

Step 6. You want the coefficient of x to be 1, so divide both sides by 2.

$$\frac{2x}{\mathbf{2}} = \frac{20}{\mathbf{2}}$$

Step 7. Simplify.

$$\boldsymbol{x = 10}$$

Step 8. Check your work by reviewing steps 1–7.

$$3\left(\frac{1}{3}x - 2\right) = 2\left(7 - \frac{1}{2}x\right)$$

$$3 \cdot \frac{1}{3}x - 3 \cdot 2 = 2 \cdot 7 - 2 \cdot \frac{1}{2}x$$

$$x - 6 = 14 - x$$

$$x - 6 + x = 14 - x + x$$

$$2x - 6 = 14$$

$$2x - 6 + 6 = 14 + 6$$

$$2x = 20$$

$$\frac{2x}{\mathbf{2}} = \frac{20}{\mathbf{2}}$$

$$x = 10$$

Step 9. Check your answer by substituting 10 for x in the original equation, $3\left(\frac{1}{3}x - 2\right) = 2\left(7 - \frac{1}{2}x\right)$.

Substitute 10 for x on the LS of the equation: $3\left(\frac{1}{3}x - 2\right) = 3\left(\frac{1}{3} \cdot 10 - 2\right) = 3\left(3\frac{1}{3} - 2\right) = 3\left(1\frac{1}{3}\right) = 3 \cdot \frac{4}{3} = 4$. Similarly, on the RS, you have $2\left(7 - \frac{1}{2}x\right) = 2\left(7 - \frac{1}{2} \cdot 10\right) = 2(7 - 5) = 2 \cdot 2 = 4$. Both sides equal 4, so 10 is the correct answer.

Translating and Solving Word Equations

Sometimes equations are expressed in words, with the word *is* (or equivalent word) indicating equality. The first step in learning to translate a word equation into mathematical symbols is to understand how addition, subtraction, multiplication, and division are expressed. Table 14.1 summarizes the most commonly used symbolism for the operations expressed as word phrases. The letter x is used in the table to represent an unknown number.

Remember that, besides indicating multiplication, parentheses are used as grouping symbols. Similarly, a fraction bar can be a grouping symbol as well as indicating division.

Table 14.1 Operational Symbolism Expressed as Word Phrases

OPERATION	SYMBOL(S) USED	EXAMPLE	SAMPLE WORD PHRASES
Addition	$+$	$x + 5$	x plus 5, the sum of x and 5, 5 added to x, 5 more than x, x increased by 5
Subtraction	$-$	$x - 3$	x minus 3, the difference between x and 3, 3 subtracted from x, 3 less than x, x decreased by 3
Multiplication	side-by-side	$\frac{1}{2}x$	$\frac{1}{2}$ times x, the product of $\frac{1}{2}$ and x, x multiplied by $\frac{1}{2}$, half of x, $\frac{1}{2}$ of x
	( )	$10(2x + 5)$	10 times the quantity $2x +$ 5, the product of 10 and the quantity $2x + 5$, the quantity $2x + 5$ multiplied by 10
Division	fraction bar	$\frac{x}{8}$	x divided by 8, the quotient of x and 8, the ratio of x to 8

For products like $10(2x + 5)$, it's important to use the phrase "the quantity $2x + 5$" when referring to the expression inside the parentheses, so that it is clear that you mean the product of 10 and the entire expression.

Problem Translate the word equation into mathematical symbols and then solve it.

a. Fourteen plus five times a number x is four. Find x.

b. Four less than three times a number x is the product of two and the quantity x plus one. Find x.

Solution

a. Fourteen plus five times a number x is four. Find x.

Step 1. Replace the word *is* with the = symbol.

Fourteen plus five times a number x = four

Step 2. Translate the left side of the word equation into mathematical symbols.

"Fourteen plus five times a number x" is $14 + 5x$

Step 3. Translate the right side of the word equation into mathematical symbols.

"four" is 4

Step 4. Write the symbolic equation and then solve for x.

$14 + 5x = 4$

$14 + 5x - 14 = 4 - 14$

$5x = -10$

$\frac{5x}{5} = \frac{-10}{5}$

$x = -2$

Step 5. Check your answer by substituting -2 for x in the word equation, "Fourteen plus five times a number x is four." Substitute -2 for x on the LS of the word equation: "Fourteen plus five times a number x" = $14 + 5(-2) = 14 + -10 = 4$. Similarly, on the RS, "four" = 4. Both sides equal 4, so -2 is the correct answer.

b. Four less than three times a number x is the product of two and the quantity x plus one. Find x.

Step 1. Replace the word *is* with the = symbol.

Four less than three times a number x = the product of two and the quantity x plus one

Step 2. Translate the left side of the word equation into mathematical symbols.

"Four less than three times a number x" is $3x - 4$

> Do not make the mistake of translating "Four less than three times a number x" as $4 - 3x$.

Step 3. Translate the right side of the word equation into mathematical symbols.

"The product of two and the quantity x plus one" is $2(x + 1)$

> "The product of two and the quantity x plus one" is not $2x + 1$. The word *quantity* means you must enclose $x + 1$ in parentheses.

Step 4. Write the symbolic equation and then solve for x.

$3x - 4 = 2(x + 1)$

$3x - 4 = 2 \cdot x + 2 \cdot 1$

$3x - 4 = 2x + 2$

$3x - 4 - 2x = 2x + 2 - 2x$

$x - 4 = 2$

$x - 4 + 4 = 2 + 4$

$x = 6$

Step 5. Check your answer by substituting 6 for x in the verbal equation, "Four less than three times a number x is the product of two and the quantity x plus one." Substitute 6 for x on the LS of the word equation: "Four less than three times a number x" = 4 less than $3(6) = 4$ less than $18 = 14$. Similarly, on the RS, "the product of two and the quantity x plus one" = the product of 2 and the quantity $(6 + 1)$ = the product of 2 and $7 = 2(7) = 14$. Both sides equal 14, so 6 is the correct answer.

Writing Equations to Solve Percent Problems

In Chapter 9, you solved percent problems using proportion concepts. You also can solve percent problems by writing and solving equations. The relationship among the elements of a percent problem is given by the formula

$$P = RB$$

where R is the *rate*, B is the *base*, and P is the *percentage*. The secret to solving percent problems is being able to identify the three elements correctly.

Start with R and B because they are usually easier to find. R is the percent in the problem and will have a % symbol or the word *percent* attached. B is the whole on which the rate is based and very often is the amount that immediately follows *% of* or *percent of*. The *percentage* is the portion of B that is determined by R. It is the other amount in the problem, and when the problem statement contains the word *is*, P is near *is*. The value of two elements will be given in the problem, and you will be solving for the third element. After you identify the three elements, write an equation using the formula $P = RB$ and then solve it.

Problem Write an equation and then solve it.

a. What is 25% of 70?

b. Fifty is 40% of what number?

c. What percent of 150 is 30?

d. A student scored 60 out of 80 questions. What is the student's percent grade?

e. A toy is on sale for \$13. This sale price is 80% of the regular price of the toy. What is the regular price?

Solution

a. What is 25% of 70?

Remember that in computations, percents must be in decimal or fraction form.

Step 1. Identify R, B, and P.

$R = 25\% = 0.25,\ B = 70,\ P = ?$

Step 2. Write an equation and then solve it.

$P = RB$

$P = (0.25)(70)$

$P = 17.5$

Step 3. State the answer.

$\underline{17.5}$ is 25% of 70.

b. Fifty is 40% of what number?

Step 1. Identify R, B, and P.

$R = 40\% = 0.4,\ B = ?,\ P = 50$

Step 2. Write an equation and then solve it.

$P = RB$

$50 = (0.4)B$

$(0.4)B = 50$ *Note:* For convenience, switch sides, so that the variable B is on the LS.

$0.4B = 50$

$\frac{50}{0.4} = 50 \div 0.4$ on your calculator. The fraction bar indicates division.

$\frac{0.4B}{0.4} = \frac{50}{0.4}$

$B = 125$

Step 3. State the answer.

50 is 40% of <u>125</u>.

c. What percent of 150 is 30?

Step 1. Identify R, B, and P.

$R = ?, B = 150, P = 30$

Step 2. Write an equation and then solve it.

$P = RB$

$30 = R(150)$

$R(150) = 30$

$R(150) = 150R$ because multiplication is commutative.

$150R = 30$

$\frac{150R}{150} = \frac{30}{150}$

$R = 0.2$

Step 3. Change 0.2 to a percent.

$R = 0.2 = 20\%$

Step 4. State the answer.

30 is <u>20%</u> of 150.

d. A student scored 60 out of 80 questions. What is the student's percent grade?

Step 1. Identify R, B, and P.

$R = ?, B = 80, P = 60$

Step 2. Write an equation and then solve it.

$P = RB$

$60 = R(80)$

$R(80) = 60$

$80R = 60$

$\frac{80R}{80} = \frac{60}{80}$

$R = 0.75$

Step 3. Change 0.75 to a percent.

$R = 0.75 = 75\%$

Step 4. State the answer.

The student's percent grade is <u>75%</u>.

e. A toy is on sale for \$13. This sale price is 80% of the regular price of the toy. What is the regular price?

Step 1. Identify R, B, and P.

$R = 80\% = 0.8$, $B = ?$, $P = \$13$

Step 2. Write an equation and then solve it.

$P = RB$

$\$13 = (0.8)B$

$(0.8)B = \$13$

$0.8B = \$13$

$\frac{0.8B}{0.8} = \frac{\$13}{0.8}$

$B = \$16.25$

Step 3. State the answer.

The regular price of the toy is <u>\$16.25</u>.

Exercise 14

For 1–4, solve the equation for x.

1. $x - 7 = 11$
2. $6x - 3 = 13$
3. $x + 3(x - 2) = 2x - 4$
4. $\frac{x+3}{5} = \frac{x-1}{2}$
5. Translate the word equation into a symbolic equation and then solve for x: Two more than three times a number x is four less than six times the number x.

For 6–10, write an equation and then solve it.

6. What is 35% of 500?
7. Ninety is 60% of what number?
8. What percent of \$144 is \$21.60?
9. A student scored 70 out of 80 questions. What is the student's percent grade?
10. A shirt is on sale for \$76. This sale price is 80% of the regular price of the shirt. What is the regular price?

15

Informal Geometry

In this chapter, you learn informal geometry concepts.

Congruence

Congruent figures have exactly the same size and same shape. They will fit exactly on top of each other.

Problem Do the two figures appear to be congruent? Yes or no?

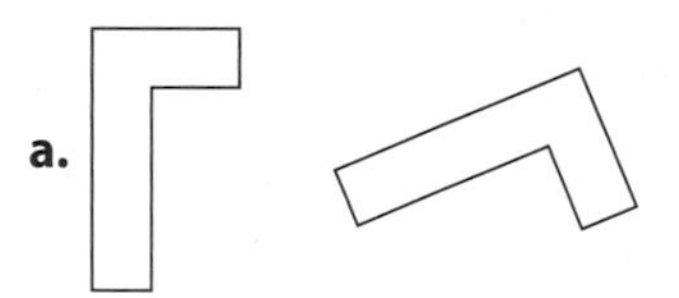

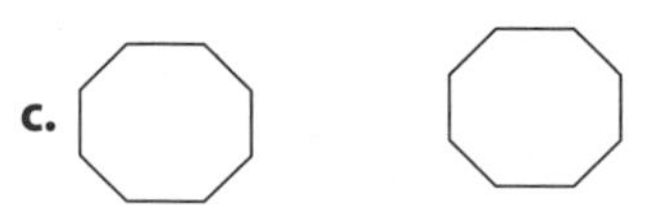

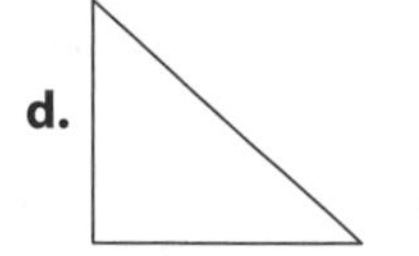

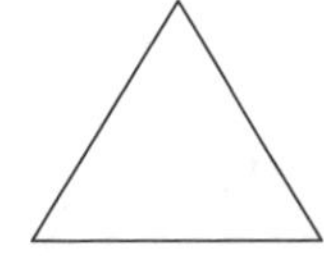

e. 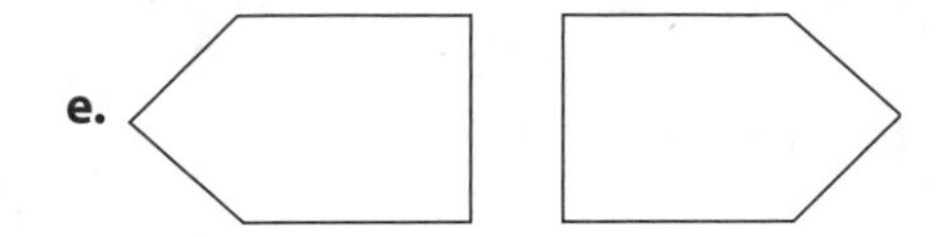

Solution

a. 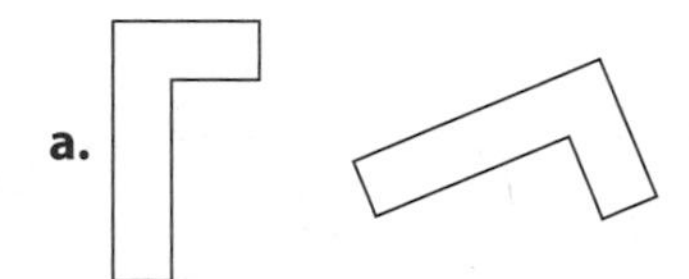

Step 1. Check whether the two figures appear to be the same size.
The two figures appear to be the same size.

Step 2. Check whether the two figures appear to have the same shape.
The two figures appear to have the same shape.

Step 3. Answer the question.
Yes, the two figures appear to be congruent.

b. 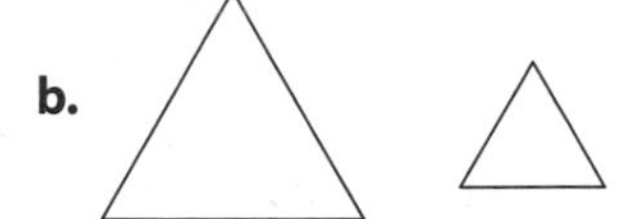

Step 1. Check whether the two figures appear to be the same size.
The two figures do not appear to be the same size.

Step 2. Answer the question.
No, the two figures do not appear to be congruent.

c. 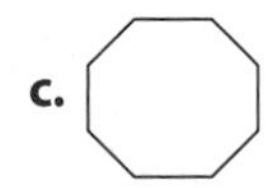 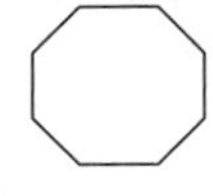

Step 1. Check whether the two figures appear to be the same size.
The two figures appear to be the same size.

Step 2. Check whether the two figures appear to have the same shape.
The two figures appear to have the same shape.

Step 3. Answer the question.

Yes, the two figures appear to be congruent.

d. 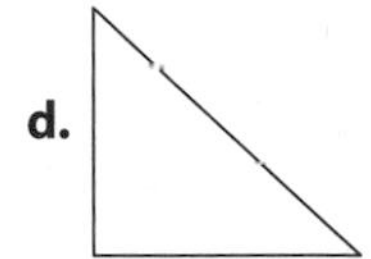 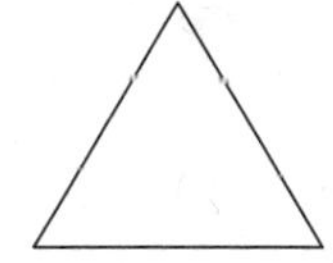

Step 1. Check whether the two figures appear to be the same size.

The two figures appear to be the same size.

Step 2. Check whether the two figures appear to have the same shape.

The two figures do not appear to have the same shape. The figure on the right appears to have all sides the same length, but the figure on the left appears to have sides that are unequal in length.

Step 3. Answer the question.

No, the two figures do not appear to be congruent.

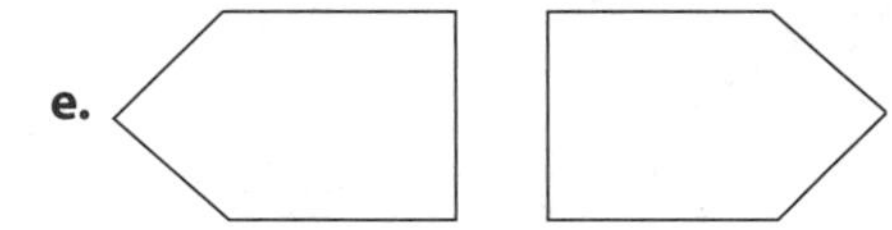

Step 1. Check whether the two figures appear to be the same size.

The two figures appear to be the same size.

Step 2. Check whether the two figures appear to have the same shape.

The two figures appear to have the same shape.

Step 3. Answer the question.

Yes, the two figures appear to be congruent.

Similarity

Similar geometric figures have the same shape, but not necessarily the same size.

All congruent figures are also similar figures. However, not all similar figures are congruent.

Problem Do the two figures appear to be similar? Yes or no?

a. 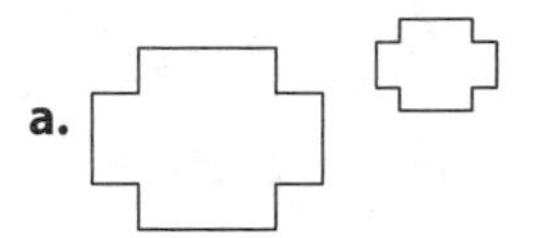

b. 

c. 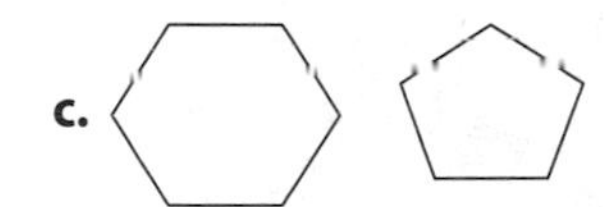

d. 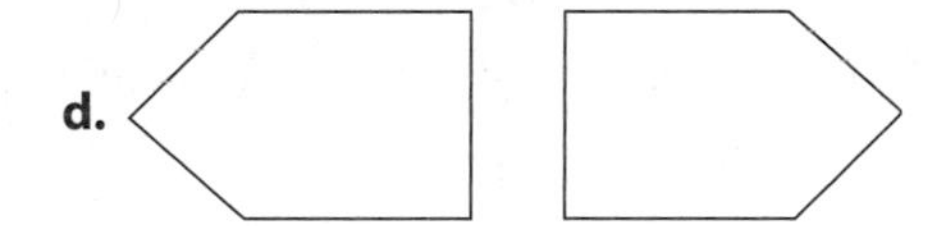

Solution

a. 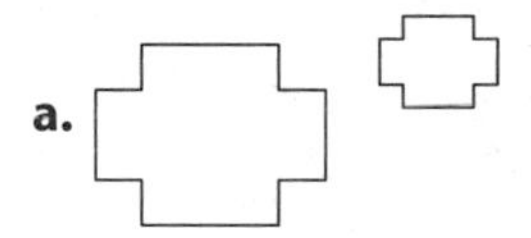

Step 1. Check whether the two figures appear to have the same shape.

The two figures appear to have the same shape.

Step 2. Answer the question.

Yes, the two figures appear to be similar.

b. 

Step 1. Check whether the two figures appear to have the same shape.

The two figures appear to have the same shape.

Step 2. Answer the question.

Yes, the two figures appear to be similar.

c. 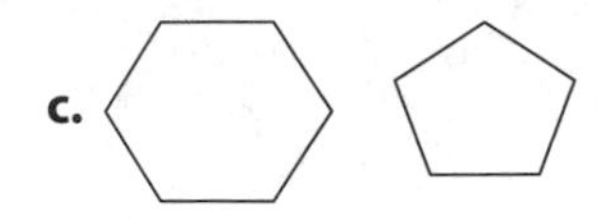

Step 1. Check whether the two figures appear to have the same shape.

The two figures do not have the same shape.

Step 2. Answer the question.

No, the two figures are not similar.

d. 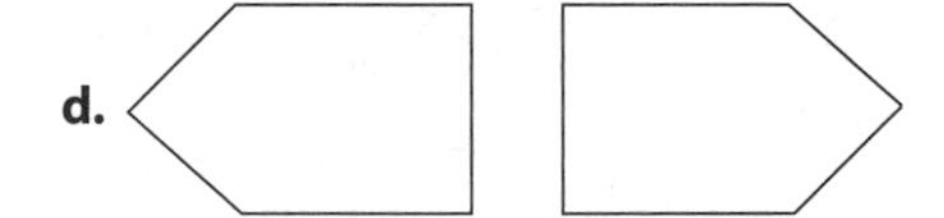

Step 1. Check whether the two figures appear to have the same shape.

The two figures appear to have the same shape.

Step 2. Answer the question.

Yes, the two figures appear to be similar.

Symmetry

Symmetry describes a characteristic of the shape of a figure or object. A figure or object has line symmetry (or is symmetric) if it can be folded exactly in half resulting in two congruent halves. The line along the fold is the *line of symmetry.*

Problem Does the figure appear to have line symmetry? Yes or no? For symmetric figures, draw a line of symmetry.

a.

b.

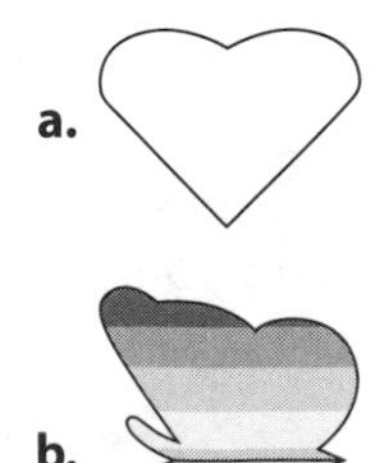

c. 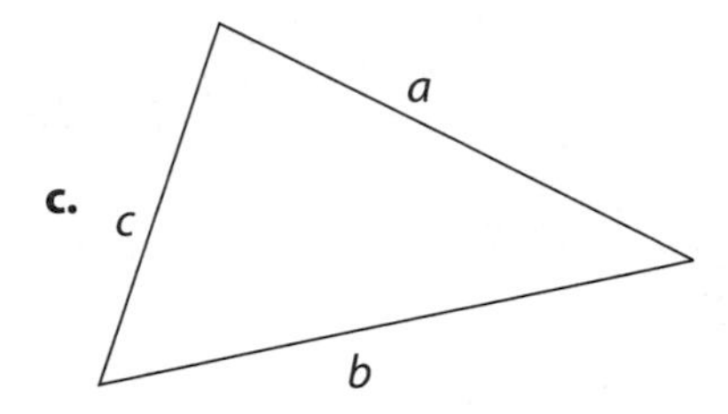

Solution

a. 

Step 1. Check whether it appears that the figure can be folded exactly in half resulting in two congruent halves.

It appears that the figure can be folded exactly in half resulting in two congruent halves.

Step 2. Answer the question.

Yes, it appears that the figure has line symmetry.

Step 3. Draw a line of symmetry for the figure.

b. 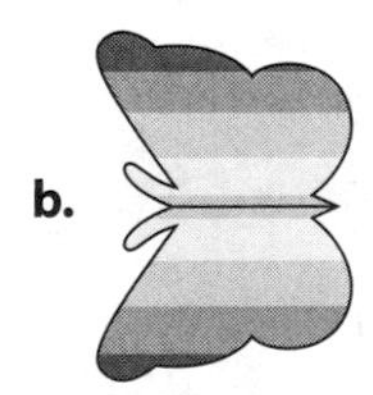

Step 1. Check whether it appears that the figure can be folded exactly in half resulting in two congruent halves.

It appears that the figure can be folded exactly in half resulting in two congruent halves.

Step 2. Answer the question.

Yes, it appears that the figure has line symmetry.

Step 3. Draw a line of symmetry for the figure.

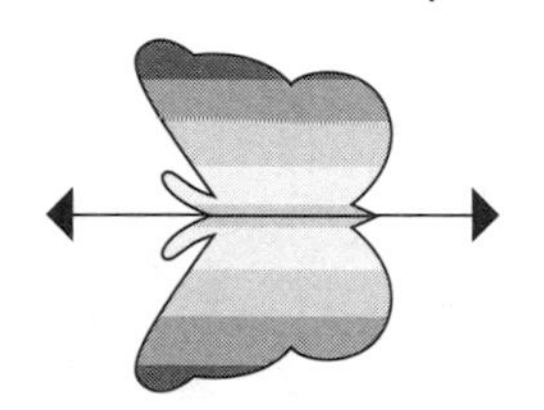

c.

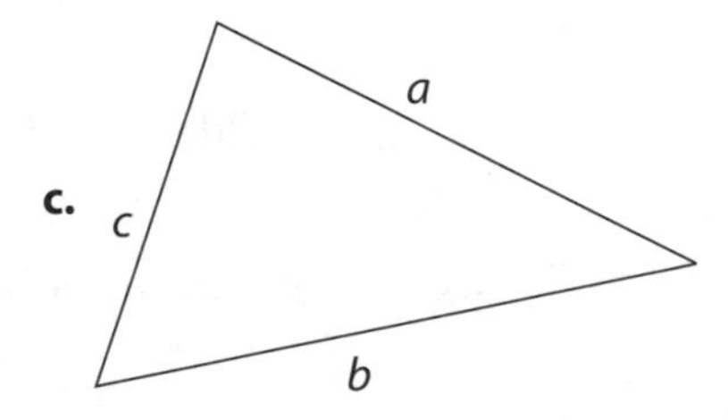

Step 1. Check whether it appears that the figure can be folded exactly in half resulting in two congruent halves.

The figure cannot be folded exactly in half resulting in two congruent halves.

Step 2. Answer the question.

No, the figure does not have line symmetry.

Some shapes have more than one line of symmetry.

Problem Draw all the lines of symmetry for the figure shown.

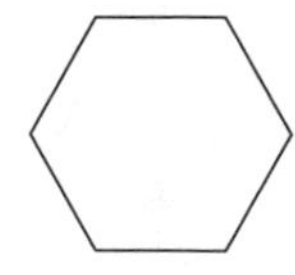

Solution

Step 1. Draw all the lines that will divide the figure into two congruent halves.

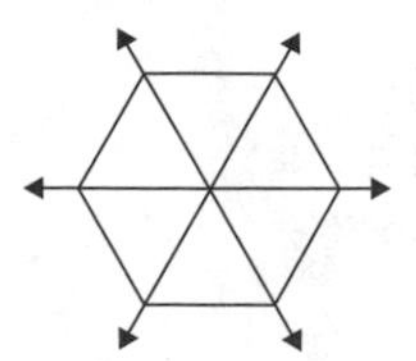

Angles

In geometry, *point*, *line*, and *plane* are basic terms. Think of a point as a location in space, a line as a set of points that goes on and on in both directions, and a plane as set of points that form a flat infinite surface.

A *ray* is a line that extends from a point. Two rays that meet at a common point form an *angle*. The *vertex* of the angle is the point where the two rays meet, as shown in Figure 15.1.

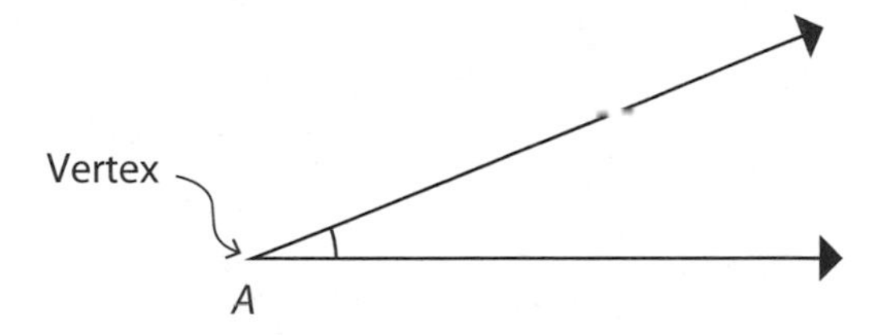

Figure 15.1 Vertex of angle *A*

You use degrees (°) to measure angles. You classify angles by the number of degrees in their measurements. An *acute angle* measures between 0° and 90°. A *right angle* measures exactly 90°. An *obtuse angle* measures between 90° and 180°. A *straight angle* measures exactly 180°.

Problem Classify the angle as an acute angle, a right angle, an obtuse angle, or a straight angle.

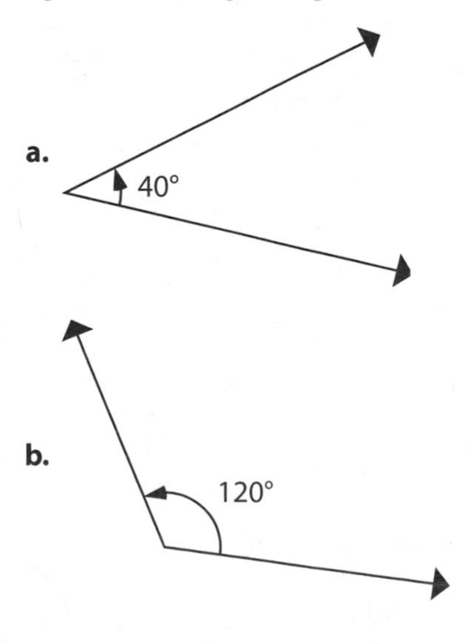

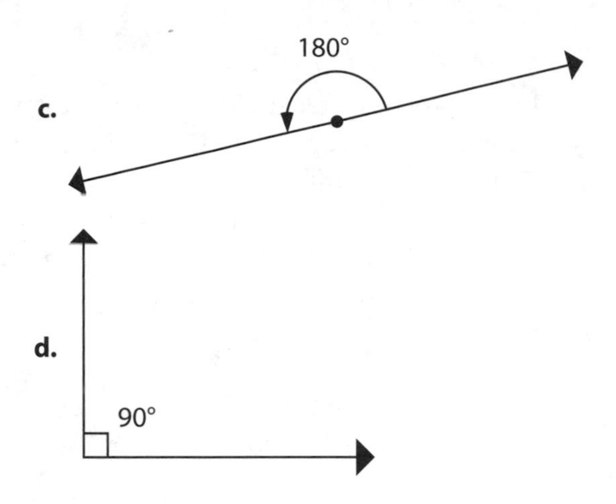

Solution

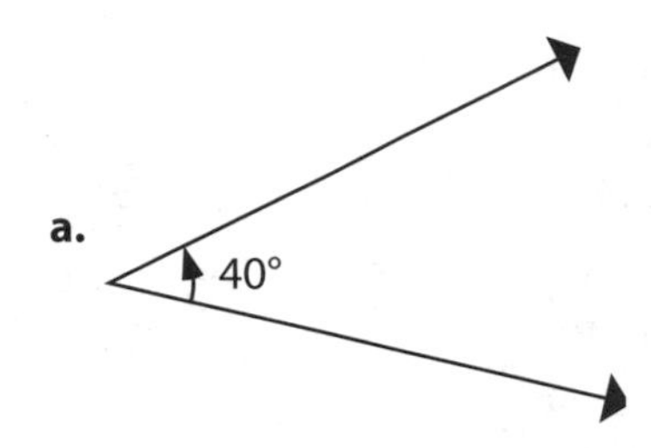

Step 1. Note the measure of the angle.

The measure of the angle is 40°.

Step 2. Classify the angle.

40° is between 0° and 90°, so the angle is acute.

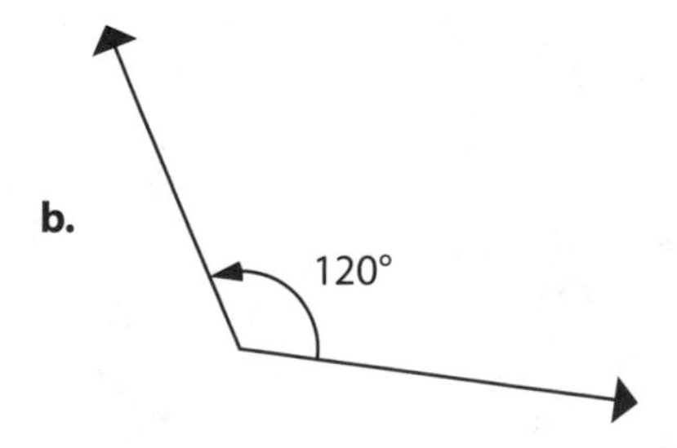

Step 1. Note the measure of the angle.

The measure of the angle is 120°.

Step 2. Classify the angle.

120° is between 0° and 180°, so the angle is obtuse.

c.

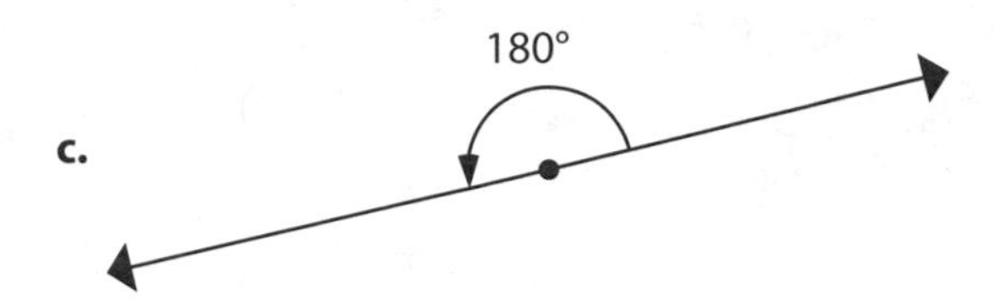

Step 1. Note the measure of the angle.

The measure of the angle is 180°.

Step 2. Classify the angle.

The angle is a straight angle.

d.

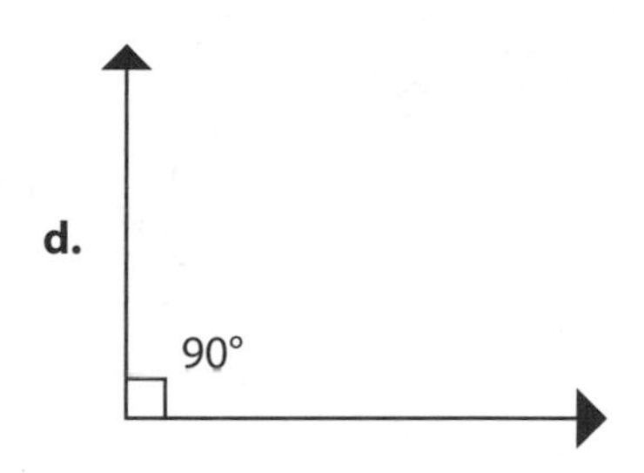

Step 1. Note the measure of the angle.

The measure of the angle is 90°.

Step 2. Classify the angle.

The angle is a right angle.

A right angle has a box as the angle indicator.

Lines

Two distinct lines in a plane can be parallel or intersecting. *Intersecting lines* cross at a common point in the plane. *Parallel lines* (in a plane) have no points in common; they never intersect. The distance between them is always the same. *Perpendicular lines* intersect at right angles.

Two lines are *distinct* if they are not the same line.

Problem State the most specific description of the two lines shown.

a.

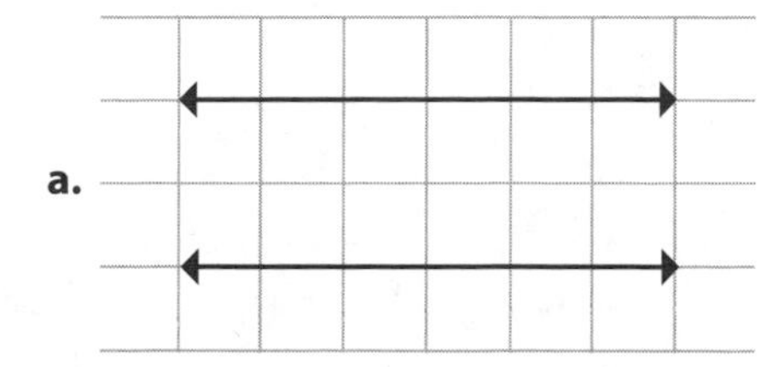

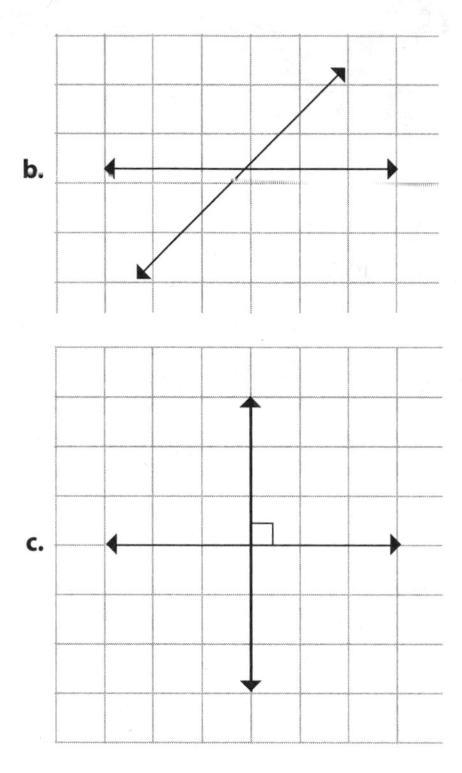

Solution

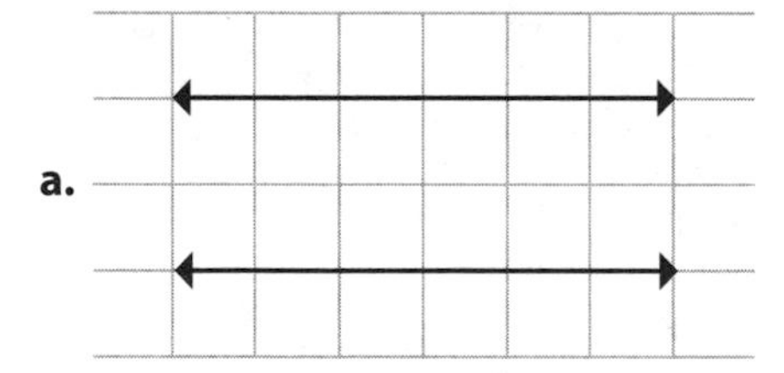

Step 1. Check whether the two lines intersect.

The two lines do not intersect.

Step 2. State the most specific description of the two lines.

The two lines are parallel.

b.

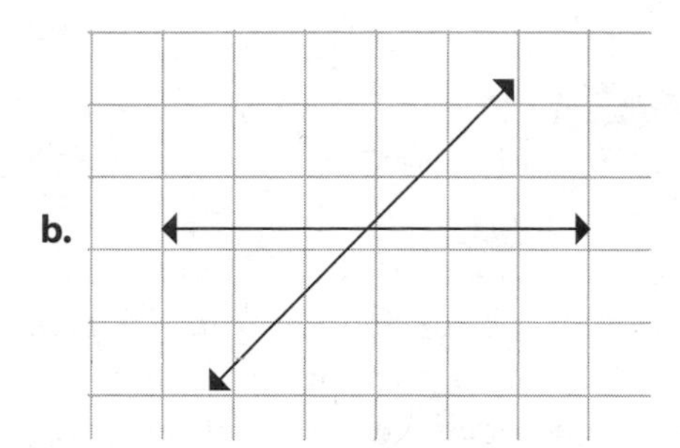

Step 1. Check whether the two lines intersect.

The two lines do intersect.

Step 2. Check whether the two lines intersect at a right angle.

The two lines do not intersect at a right angle.

Step 3. State the most specific description of the two lines.

The two lines are intersecting lines.

c.

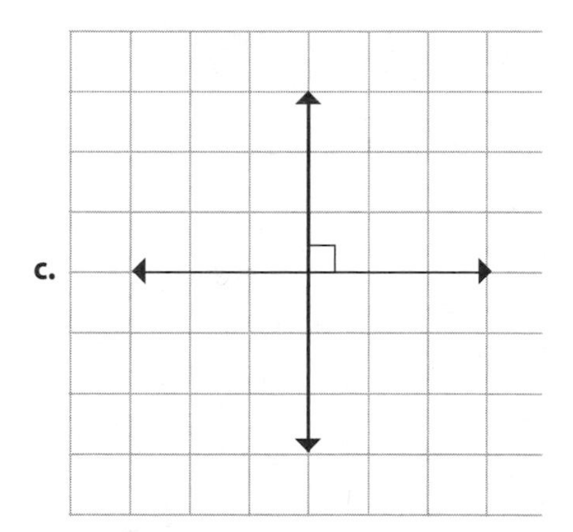

Step 1. Check whether the two lines intersect.

The two lines do intersect.

Step 2. Check whether the two lines intersect at a right angle.

The two lines do intersect at a right angle.

Step 3. State the most specific description of the two lines.

The two lines are perpendicular.

Polygons

A *polygon* is a simple, closed figure in a plane composed of *sides* that are straight line segments, fitted end to end. The sides meet only at their end points, and no two sides with a common endpoint lie on the same straight line. Polygons are named by the number of sides they have. A *triangle* is a three-sided polygon. A *quadrilateral* is a four-sided polygon. A *pentagon* is a five-sided polygon. A *hexagon* is a six-sided polygon. A *heptagon* is a seven-sided polygon. An *octagon* is an eight-sided polygon. Other polygons with additional sides have special names as well. However, eventually, at a high number of sides, you simply speak of the polygon as an *n-gon*. If all the sides of a polygon are congruent, then the polygon is a *regular polygon*.

Problem Name the polygon shown.

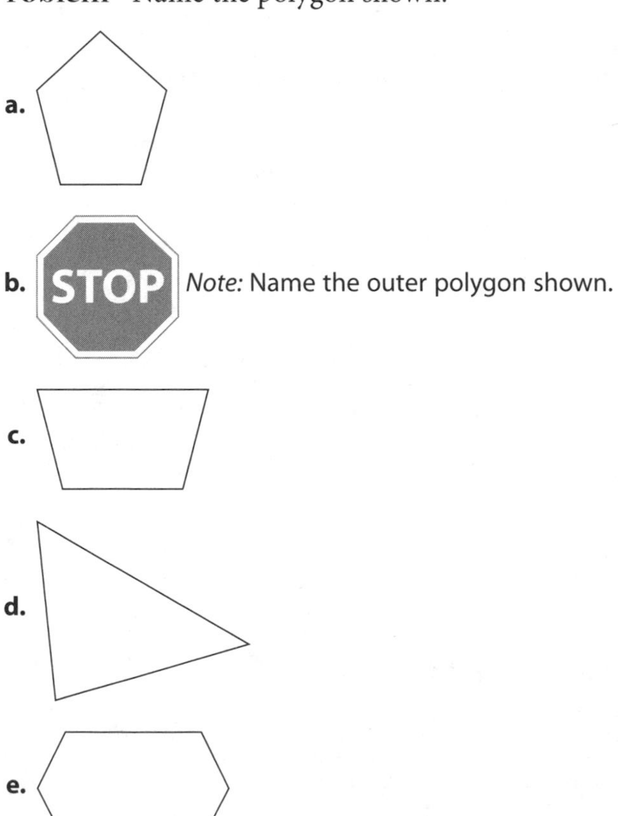

Solution

a.

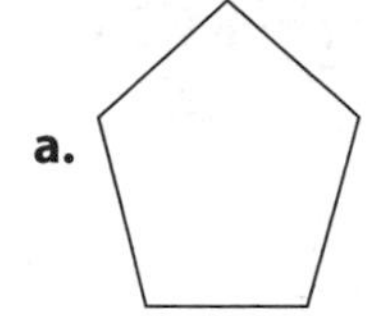

Step 1. Count the number of sides.

5

Step 2. Name the polygon.

The polygon has five sides, so it is a pentagon.

b.

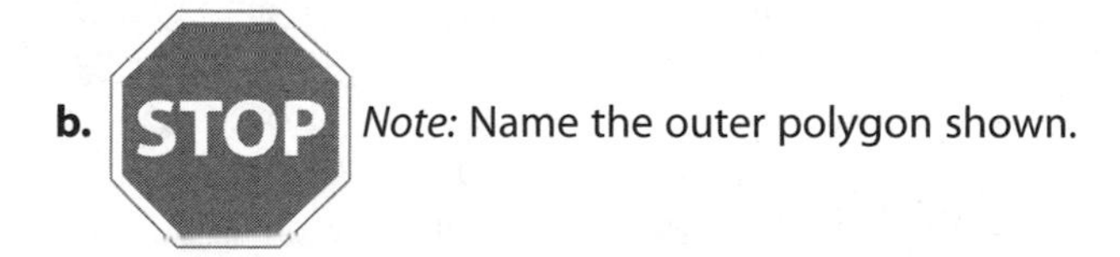

Note: Name the outer polygon shown.

Step 1. Count the number of sides.

8

Step 2. Name the polygon.

The polygon has eight sides, so it is an octagon.

c.

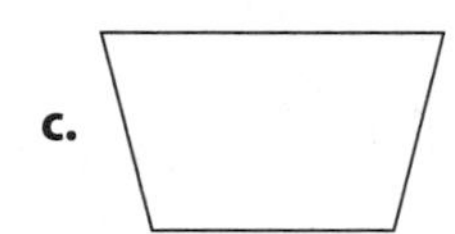

Step 1. Count the number of sides.

4

Step 2. Name the polygon.

The polygon has four sides, so it is a quadrilateral.

d. 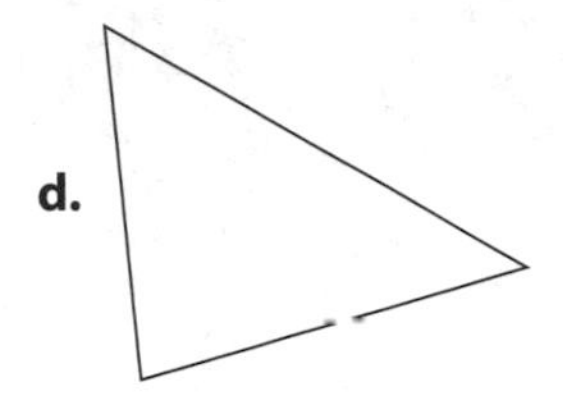

Step 1. Count the number of sides.

3

Step 2. Name the polygon.

The polygon has three sides, so it is a triangle.

e. 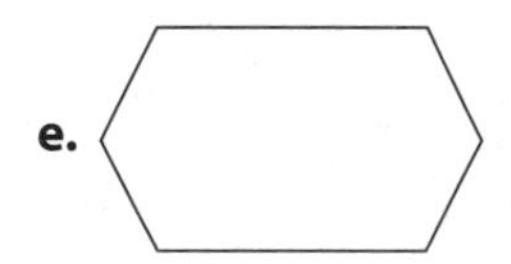

Step 1. Count the number of sides.

6

Step 2. Name the polygon.

The polygon has six sides, so it is a hexagon.

Triangles

Triangles are classified in two different ways. You classify triangles according to their sides as equilateral, isosceles, or scalene. An *equilateral triangle* has three congruent sides. An *isosceles triangle* has at least two congruent sides. A *scalene triangle* has no congruent sides.

All equilateral triangles are isosceles triangles. However, not all isosceles triangles are equilateral triangles.

Problem State the most specific name of the triangle according to its sides. (*Note:* Sides labeled with the same letter are congruent.)

a. 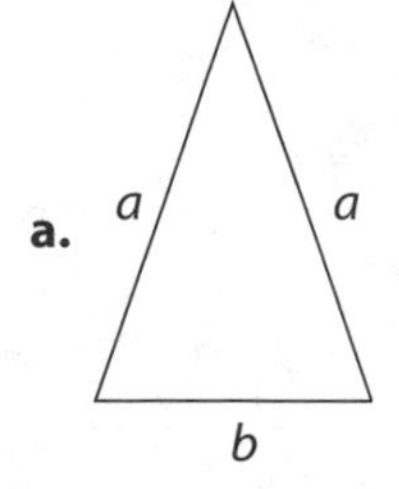

b. 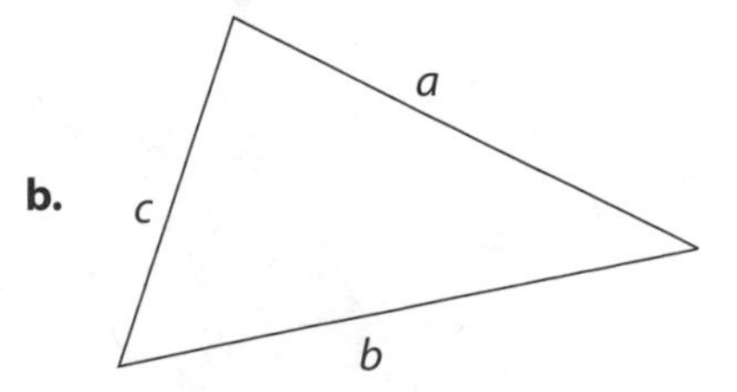

c. 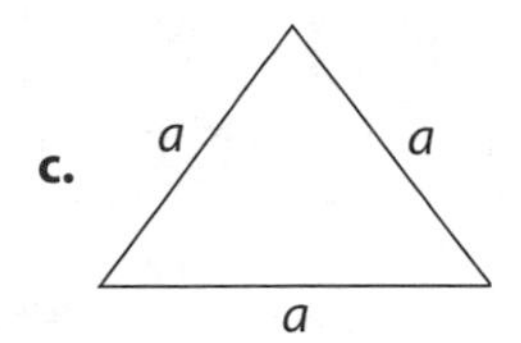

Solution

a. 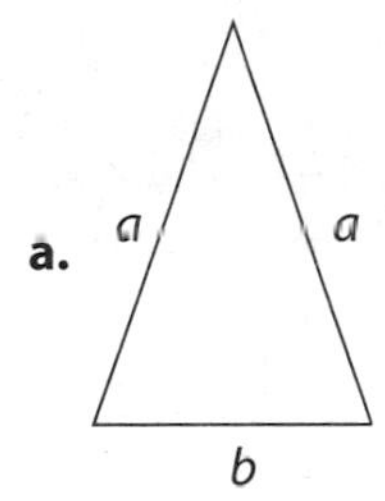

Step 1. Count the number of congruent sides.

2

Step 2. State the most specific name of the triangle.

The triangle has two congruent sides, so it is an isosceles triangle.

b. 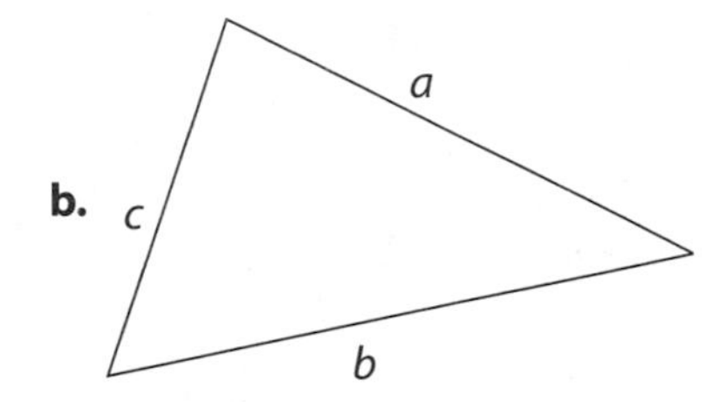

Step 1. Count the number of congruent sides.

0

Step 2. State the most specific name of the triangle.

The triangle has no congruent sides, so it is a scalene triangle.

c. 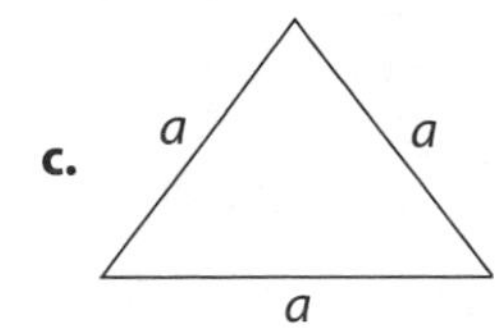

Step 1. Count the number of congruent sides.

3

Step 2. State the most specific name of the triangle.

The triangle has three congruent sides, so it is an equilateral triangle.

Another way to classify triangles is according to their interior angles. The sum of the interior angles of a triangle is 180°. An *acute triangle* has three acute angles. A *right triangle* has exactly one right angle. An *obtuse triangle* has exactly one obtuse angle.

> Because the sum of the angles of a triangle is 180°, the other two angles of a right triangle are acute angles.

> Because the sum of the angles of a triangle is 180°, the other two angles of an obtuse triangle are acute angles.

Problem Name the triangle according to its angles.

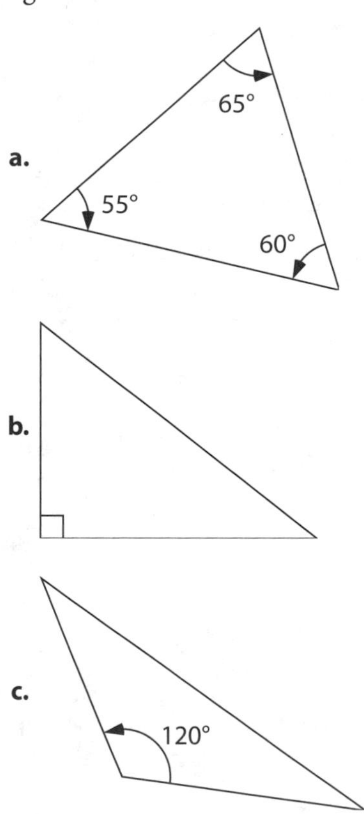

Solution

a.

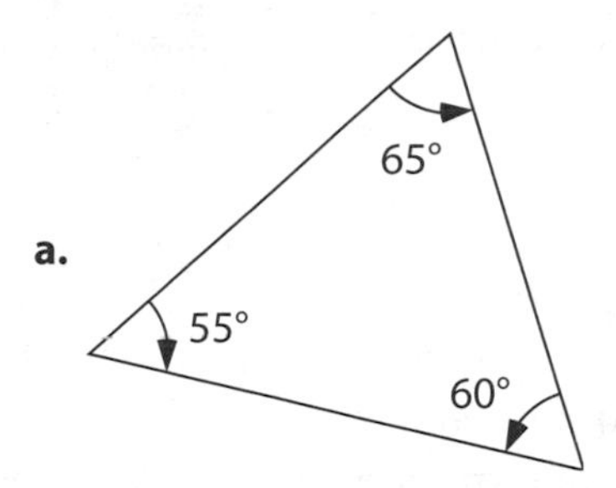

Step 1. Describe the angles of the triangle.

All three angles of the triangle are acute.

Step 2. Name the triangle according to its angles.

The triangle has three acute angles, so the triangle is an acute triangle.

b.

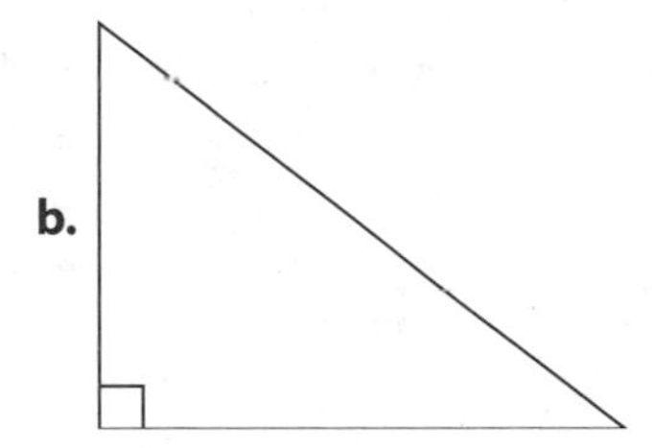

Step 1. Describe the angles of the triangle.

There is one right angle and two acute angles.

Step 2. Name the triangle according to its angles.

The triangle has exactly one right angle, so the triangle is a right triangle.

c.

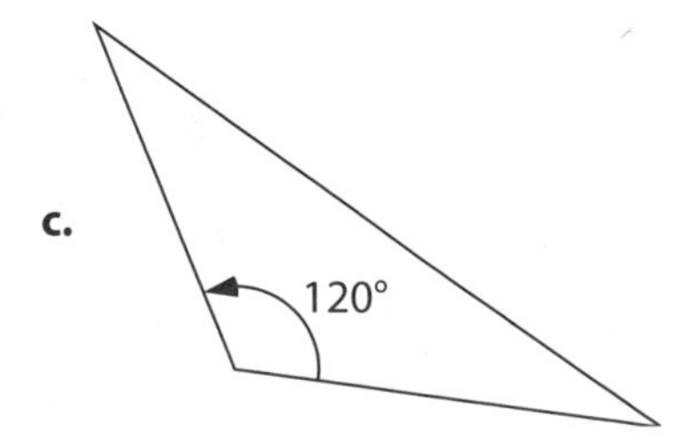

Step 1. Describe the angles of the triangle.

There is one obtuse angle and two acute angles.

Step 2. Name the triangle according to its angles.

The triangle has exactly one obtuse angle, so the triangle is an obtuse triangle.

Quadrilaterals

A *quadrilateral* is a four-sided polygon. Two special kinds of quadrilaterals are trapezoids and parallelograms. A *trapezoid* has exactly one pair of parallel sides. In a *parallelogram*, opposite sides are parallel and congruent. *Note:* Some texts define a trapezoid as a quadrilateral that has *at least* one pair of parallel sides. This situation is one of the few times that mathematicians do not agree on the definition of a term.

Some parallelograms have special names because of their special properties. A *rhombus* is a parallelogram that has four congruent sides. A *rectangle* is a parallelogram that has four interior right angles. A *square* is a parallelogram that has four interior right angles and four congruent sides. To be more specific, a *square* is a rectangle that has four congruent sides. You also can say that a *square* is a rhombus that has four interior right angles.

If a parallelogram has one interior right angle, then the other three interior angles also are right angles.

Do not make the mistake of thinking that squares are not rectangles. *All* squares are rectangles; however, not all rectangles are squares.

Problem State the most specific name of the quadrilateral. (*Note:* Sides labeled the same are congruent, and sides that look parallel are parallel.)

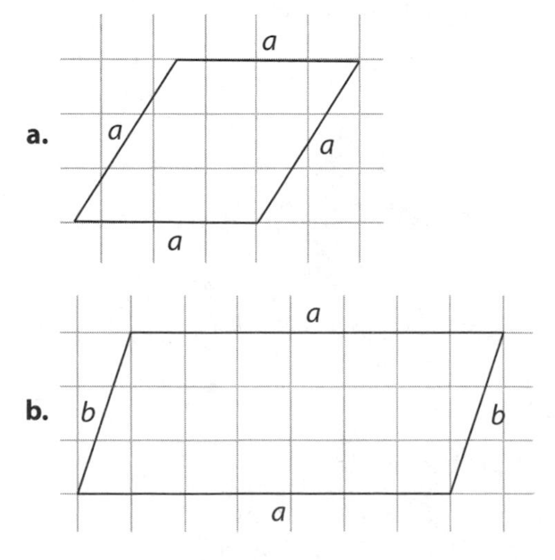

c. **d.**

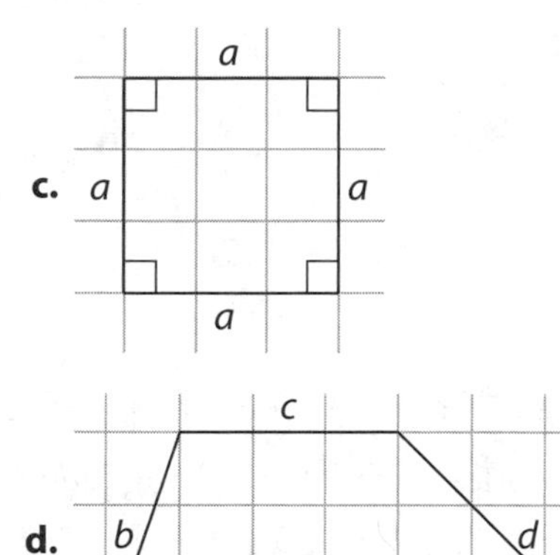

e.

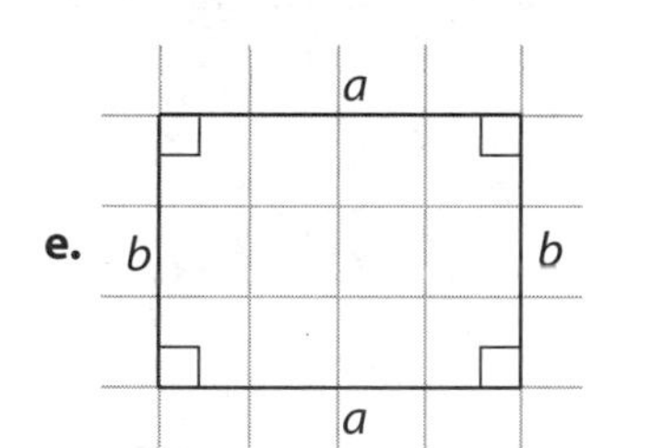

Solution

a.

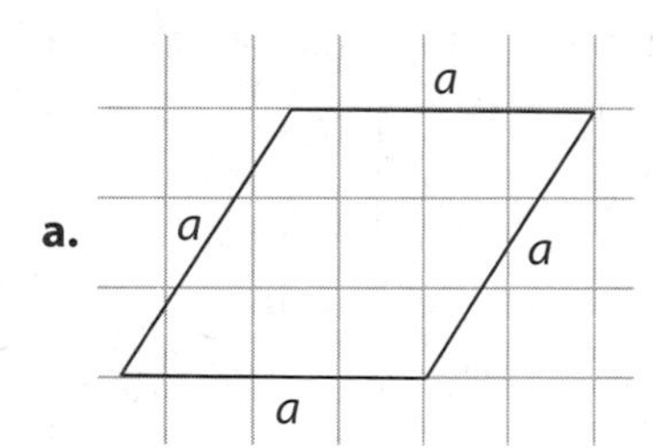

Step 1. Determine whether the quadrilateral is a trapezoid or a parallelogram.

The quadrilateral has opposite sides parallel and congruent, so it is a parallelogram.

Step 2. State the most specific name of the parallelogram.

The parallelogram has four congruent sides, so it is a rhombus.

b.

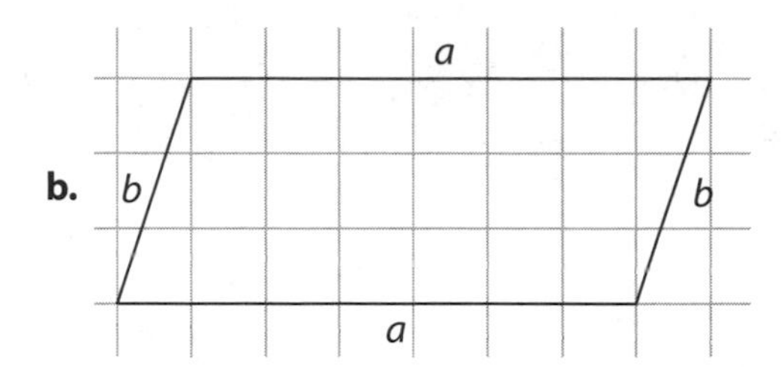

Step 1. Determine whether the quadrilateral is a trapezoid or a parallelogram.

The quadrilateral has opposite sides parallel and congruent, so it is a parallelogram.

Step 2. State the most specific name of the parallelogram.

The parallelogram does not have a special name; it's simply a parallelogram.

c.

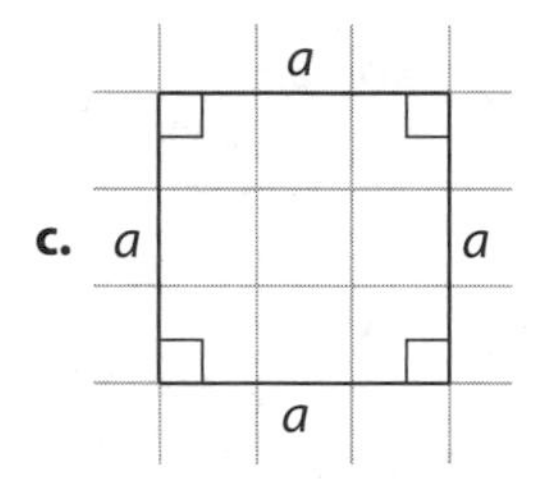

Step 1. Determine whether the quadrilateral is a trapezoid or a parallelogram.

The quadrilateral has opposite sides parallel and congruent, so it is a parallelogram.

Step 2. State the most specific name of the parallelogram.

The parallelogram has four interior right angles and four congruent sides, so it is a square.

d.

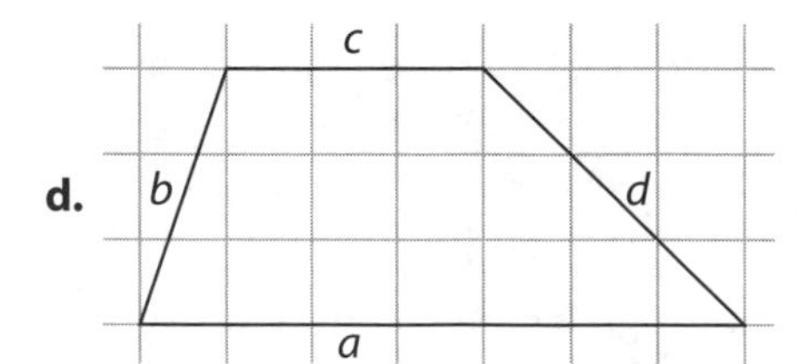

Step 1. Determine whether the quadrilateral is a trapezoid or a parallelogram.

The quadrilateral has exactly one pair of parallel sides, so it is a trapezoid.

e. 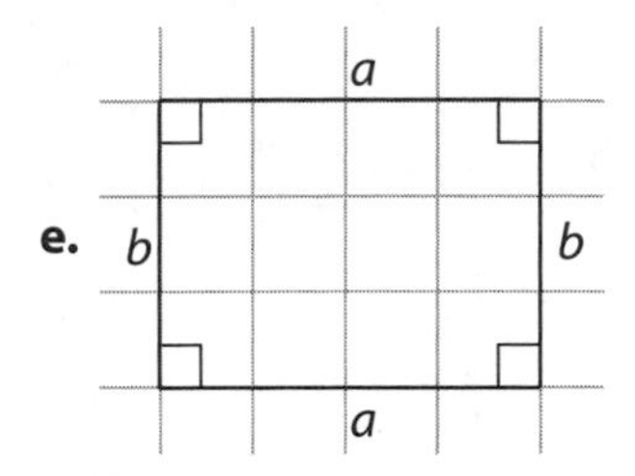

Step 1. Determine whether the quadrilateral is a trapezoid or a parallelogram.

The quadrilateral has opposite sides parallel and congruent, so it is a parallelogram.

Step 2. State the most specific name of the parallelogram.

The parallelogram has four interior right angles, so it is a rectangle.

Parts of a Circle

A *circle* is a closed figure in a plane for which all points on the figure are the same distance from a point within, called the *center.* A *radius* of a circle is a line segment joining the center of the circle to any point on the circle. The *diameter* is a line segment through the center of the circle with endpoints on the circle. The diameter of a circle is twice the radius. Conversely, the radius of a circle is half the diameter. See Figure 15.2.

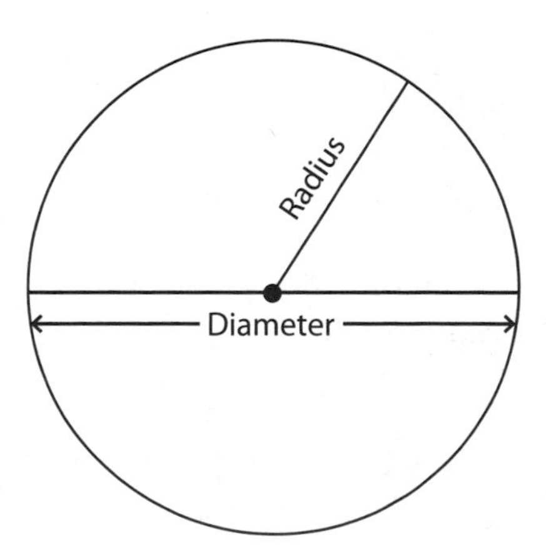

Figure 15.2 Circle

Problem Answer the question given.

a. What is the length of the diameter of the circle shown?

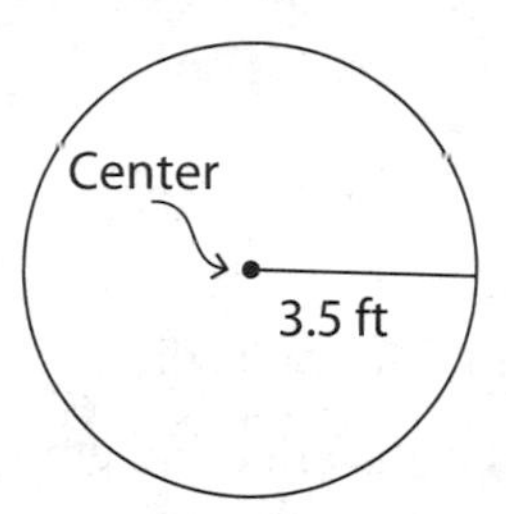

b. What is the length of the radius of the circle shown?

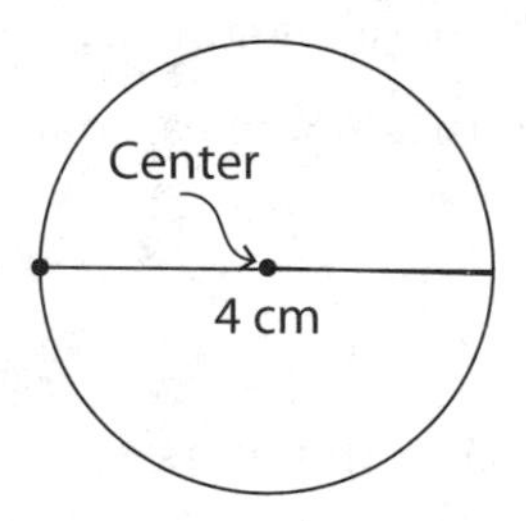

Solution

a. What is the length of the diameter of the circle shown?

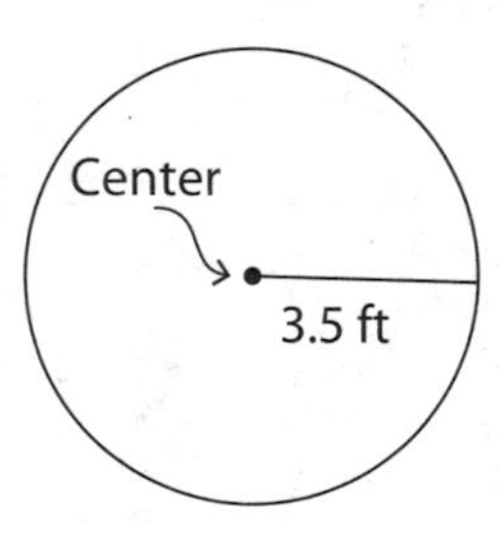

Step 1. Determine the information given.

The radius and center of the circle are shown. The radius has a length of 3.5 ft.

Step 2. Find the diameter by multiplying the length of the radius by 2.

$\text{diameter} = 2 \times 3.5 \text{ ft} = 7 \text{ ft}$

b. What is the length of the radius of the circle shown?

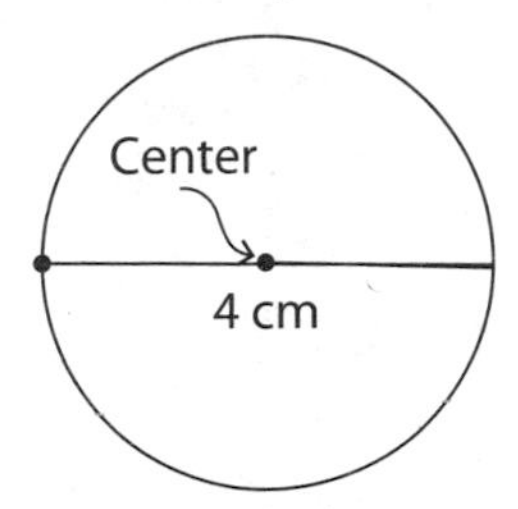

Step 1. Determine the information given.

The diameter and center of the circle are shown. The diameter has a length of 4 cm.

Step 2. Find the radius by multiplying the length of the diameter by $\frac{1}{2}$.

radius $= \frac{1}{2} \times 4 \text{ cm} = 2 \text{ cm}$

Solid Figures

Solid figures are three-dimensional figures that occupy space. The solid figures you should be able to recognize are prisms, pyramids, cylinders, cones, and spheres.

A *prism* is a solid with two congruent *faces* in parallel planes. The two congruent faces are the *bases* of the prism. The other faces of a prism are parallelograms. The bases of a prism can have the shape of any polygon. Prisms are named according to the shape of their bases. A *cube* is a special rectangular prism that has six congruent faces, all of which are squares.

A *pyramid* is a solid with exactly one base and whose sides are triangles. The base can have the shape of any polygon. Pyramids are named according to the shape of their bases.

Problem Determine whether the solid shown is a prism or a pyramid, and then state the most specific name for the solid.

a.

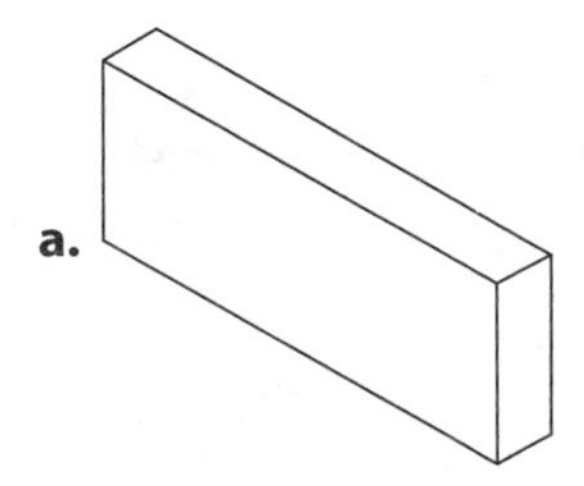

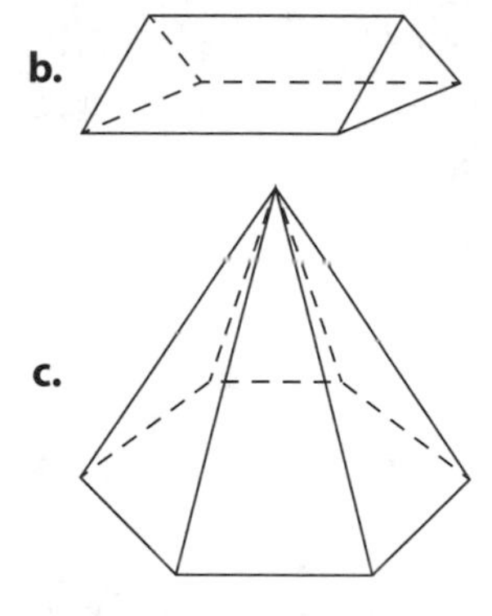

Solution

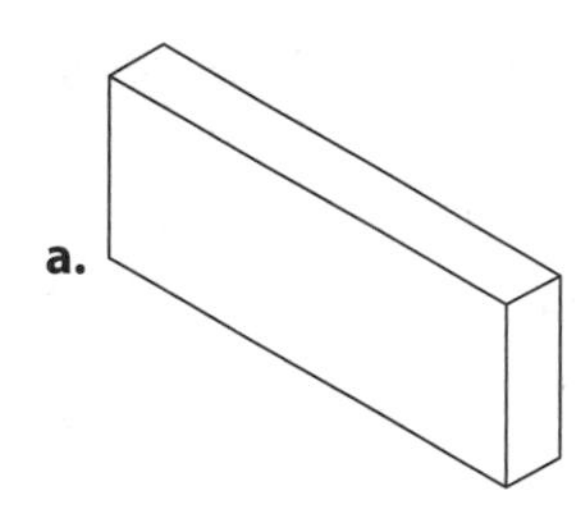

Step 1. Determine whether the solid is a prism or a pyramid.

The solid has two congruent and parallel bases, and its faces are parallelograms, so the solid is a prism.

Step 2. State the most specific name for the prism.

The prism has rectangular bases, so it is a rectangular prism.

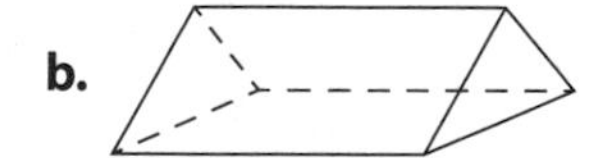

Step 1. Determine whether the solid is a prism or a pyramid.

The solid has two congruent and parallel base, and its faces are parallelograms, so the solid is a prism.

Step 2. State the most specific name for the prism.

The prism has triangular bases, so it is a triangular prism.

c.

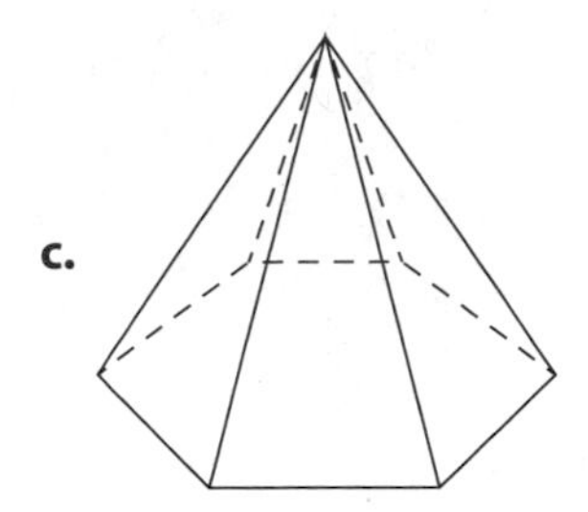

Step 1. Determine whether the solid is a prism or a pyramid.

The solid has exactly one base, and its faces are triangles, so the solid is a pyramid.

Step 2. State the most specific name for the prism.

The pyramid has a hexagonal base, so it is a hexagonal pyramid.

A *cylinder* has two parallel congruent bases, which are circles. It has one rectangular side that wraps around. A *cone* is a three-dimensional solid that has one circular base. It has a curved side that wraps around.

A *sphere* is a three-dimensional figure that is shaped like a ball. Every point on the figure is the same distance from a point within, called the *center* of the sphere. The *radius* of the sphere is a line segment from the center of the sphere to any point on the sphere. The *diameter* of the sphere is a line segment joining two points of the sphere and passing through its center. The radius of the sphere is half the diameter. Conversely, the diameter is twice the radius. See Figure 15.3.

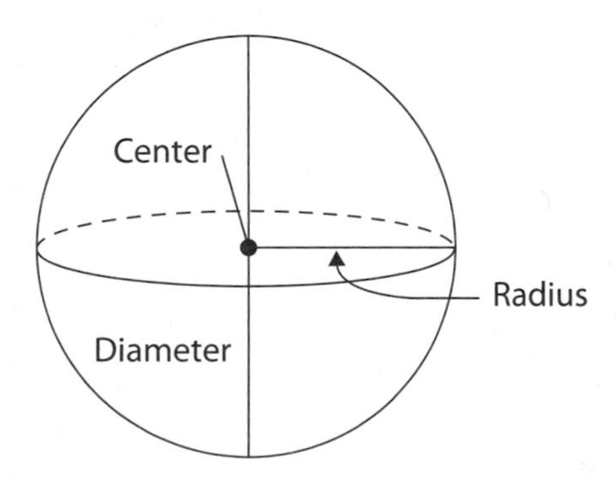

Figure 15.3 Sphere

Problem Determine whether the item shown is most like a cylinder, a cone, or a sphere.

a.

b.

c.

Solution

a.

Step 1. Determine whether the can is most like a cylinder, a cone, or a sphere.

The can has two parallel bases, which are circles. It has one rectangular side that wraps around, so it is most like a cylinder.

b. 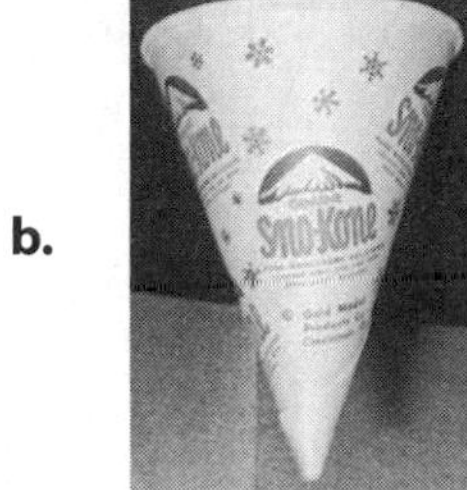

Step 1. Determine whether the snow cone cup is most like a cylinder, a cone, or a sphere.

The snow cone cup has one circular base (at the top) and a curved side that wraps around, so it is most like a cone.

c. 

Step 1. Determine whether the solid is most like a cylinder, a cone, or a sphere.

The solid has the shape of a ball, so it is a sphere.

Exercise 15

For 1–3, do the two figures appear to be congruent? Yes or no?

For 4–6, do the two figures appear to be similar? Yes or no?

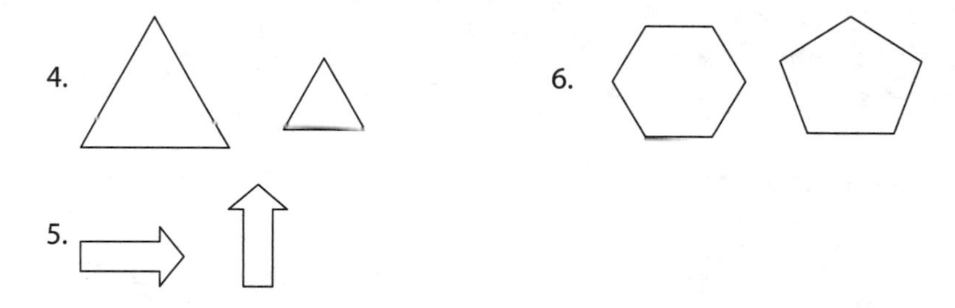

For 7 and 8, if the figure has line symmetry, draw all the lines of symmetry. If the figure does not have line symmetry, state that it does not.

7.

8.

For 9–11, classify the angle as an acute angle, a right angle, an obtuse angle, or a straight angle.

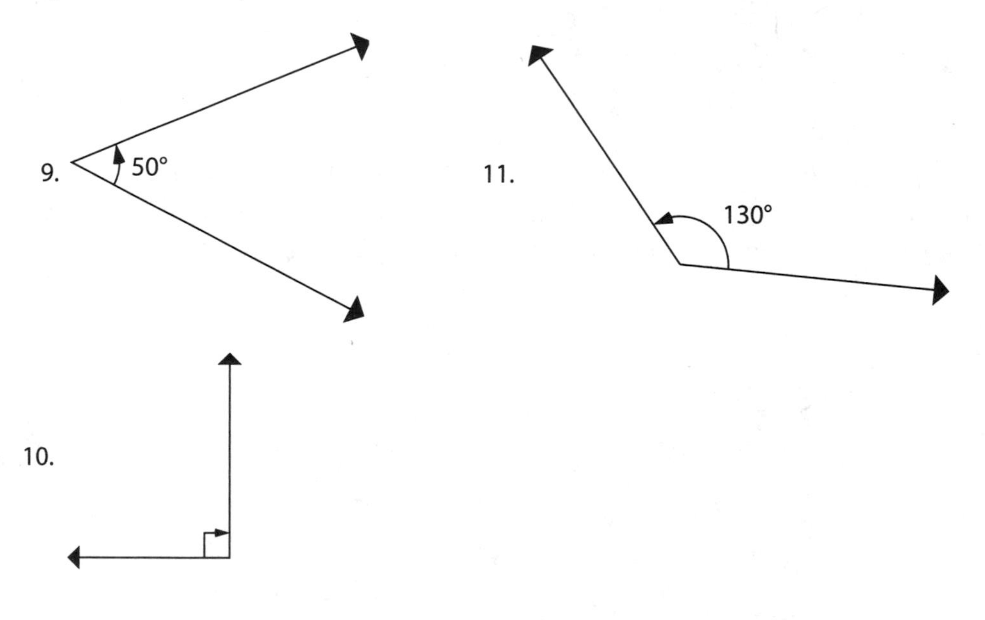

For 12 and 13, state the most specific description of the two lines shown.

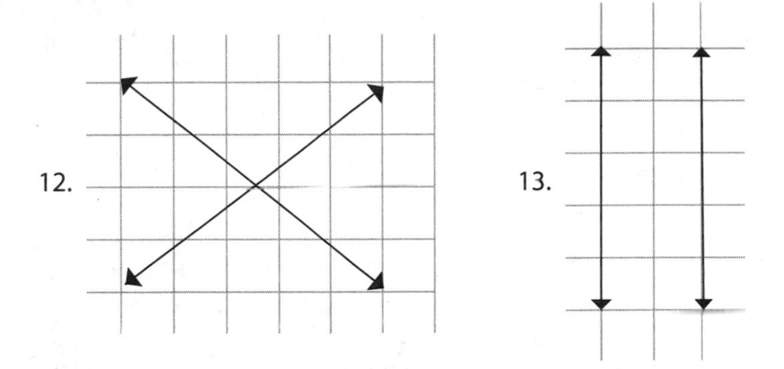

For 14 and 15, name the polygon shown.

For 16 and 17, state the most specific name of the triangle according to its sides.

For 18–20, name the triangle according to its angles.

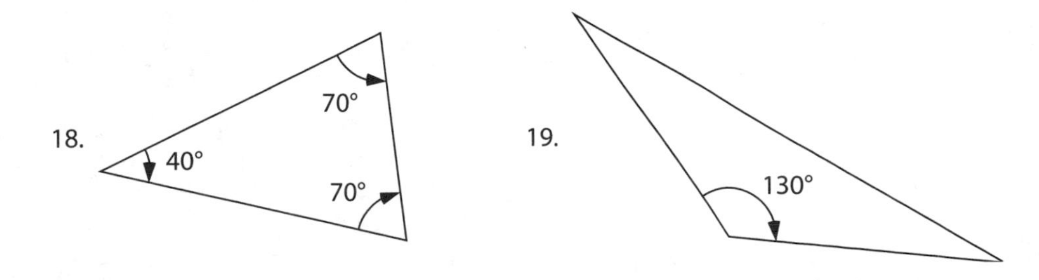

20. 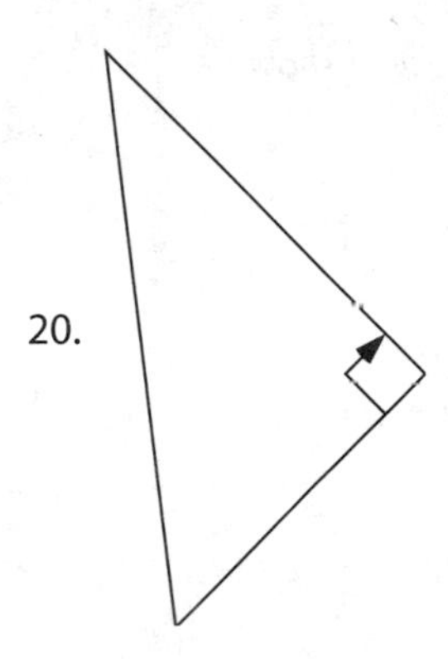

For 21–23, state the most specific name of the quadrilateral. (Note: *Sides labeled the same are congruent, and sides that look parallel are parallel.)*

21. 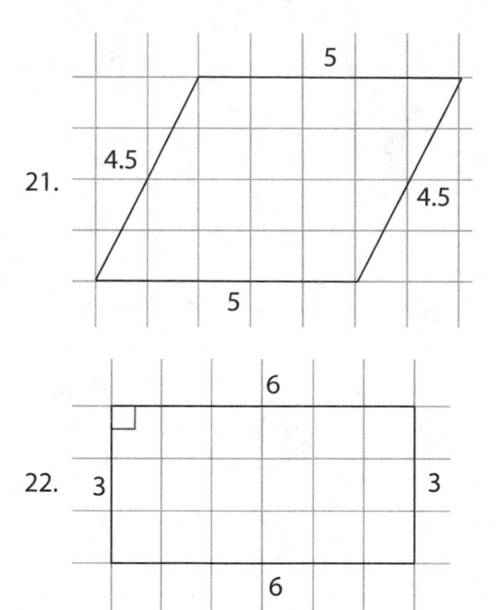

22.

23. 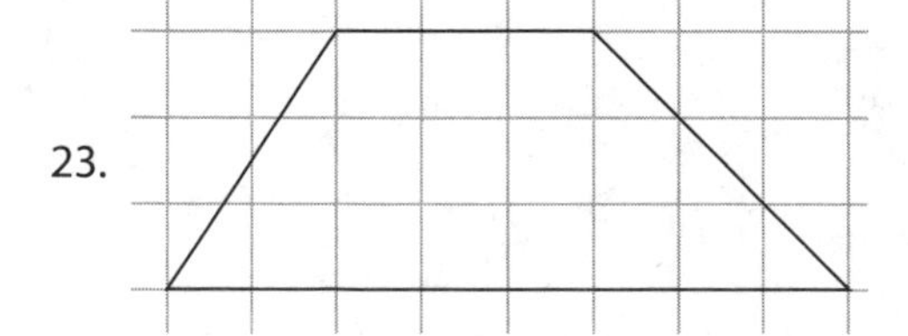

24. What is the length of the diameter of the circle shown?

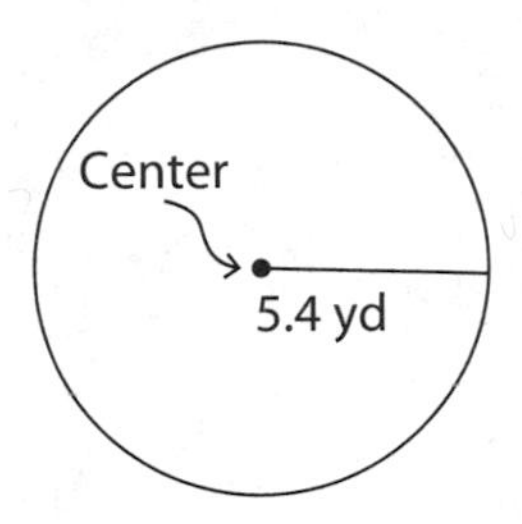

For 25–30, state the most specific name for the solid.

25.

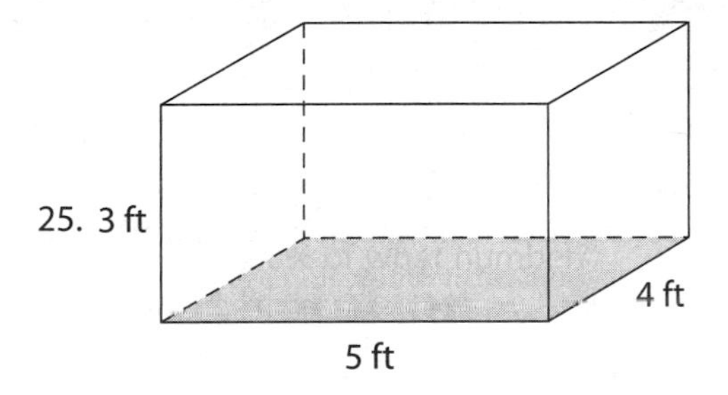

28.

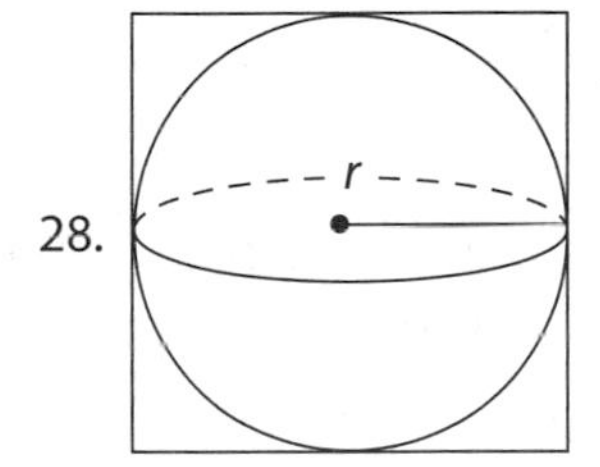

26.

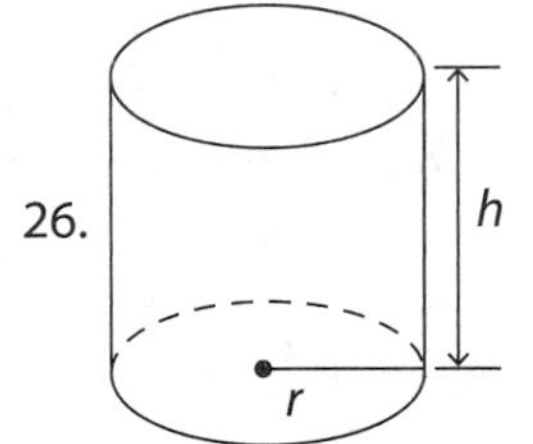

29.

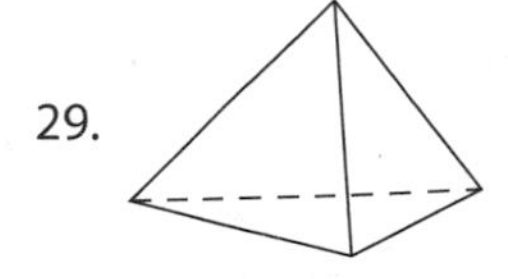

27.

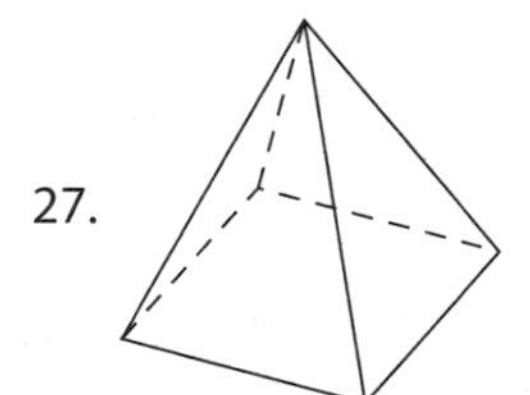

30.

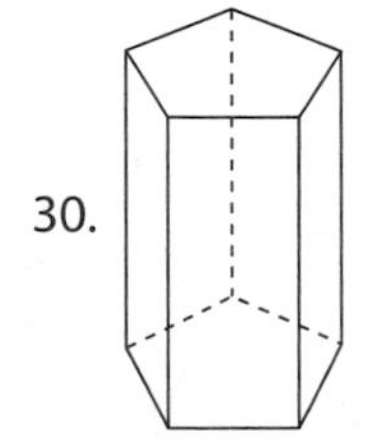

16

Perimeter, Area, and Volume

In this chapter, you learn to find the perimeter, area, and volume of geometric figures.

Perimeter

The *perimeter* of a simple, closed plane figure is a one-dimensional concept and is the distance around the figure. The perimeter is always measured in units of length such as inches, feet, centimeters, or meters. What length of fence is needed to enclose a yard? What length of decorative border is needed for a mirror? Situations such as these call for measuring the perimeter.

As an example, find the perimeter of the following figure.

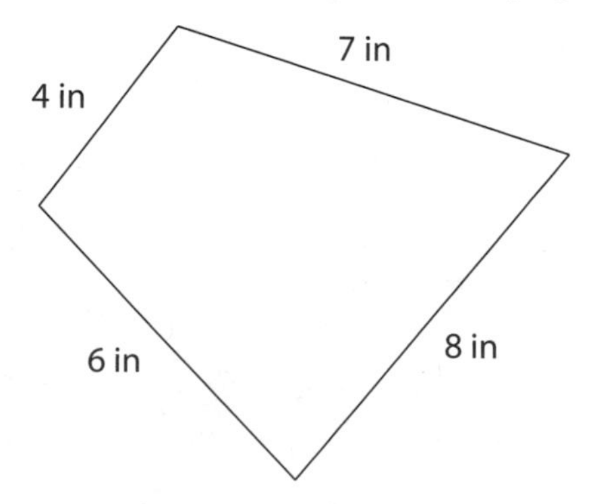

To find the perimeter, add the lengths of the four sides:

$$\text{Perimeter} = 4 \text{ in} + 7 \text{ in} + 8 \text{ in} + 6 \text{ in} = 25 \text{ in}$$

Problem Find the perimeter of the figure shown.

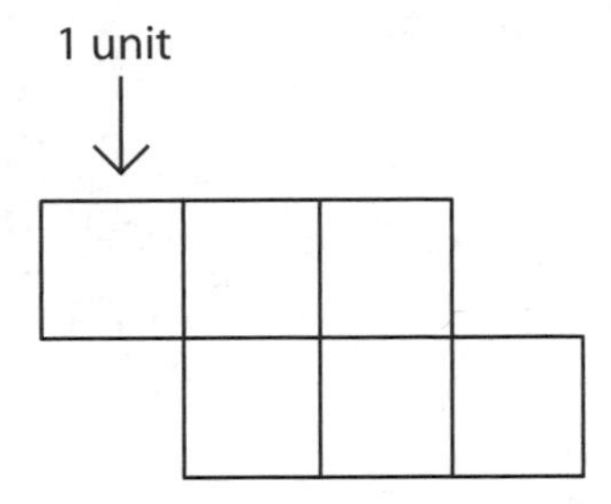

Solution The perimeter is the distance around the figure.

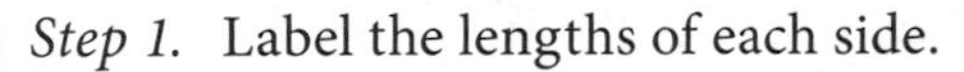

Step 1. Label the lengths of each side.

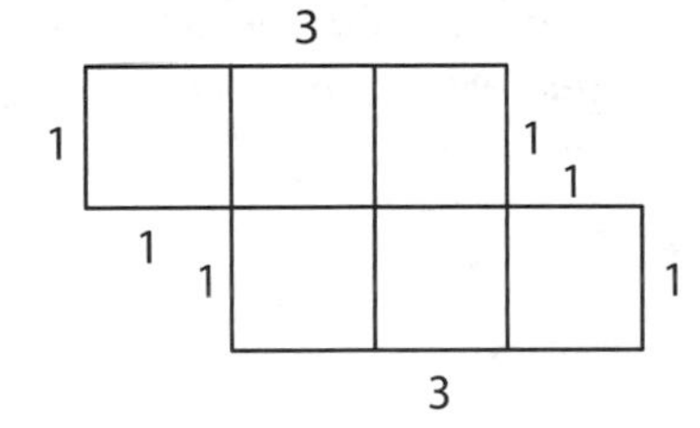

Step 2. Sum the lengths of the sides to get the perimeter.

Perimeter $= 3 + 1 + 1 + 1 + 3 + 1 + 1 + 1 = 12$ units

Perimeter of Special Shapes

Some polygons have characteristics that enable you to deduce formulas for their perimeters.

A *rectangle* has opposite sides that are congruent. If the length is denoted by l and the width by w, then the formula for the perimeter P is $P = 2l + 2w$.

$P = 2l + 2w$

w

l

A *square* has four congruent sides. If the side lengths are denoted by s, then the formula for the perimeter is $P = 4s$.

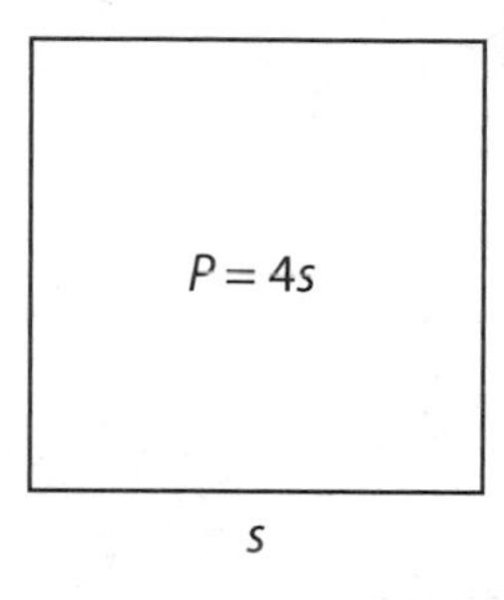

The important thing to remember is not the formulas but that the perimeter is the distance around the figure. Formulas are simply shortcuts to use when applicable.

Several other figures often encountered in the study of geometry include the following.

Recall from Chapter 15 that a *triangle* is a simple, closed plane figure with three sides. The perimeter of a triangle is the sum of its sides; that is, $P = a + b + c$.

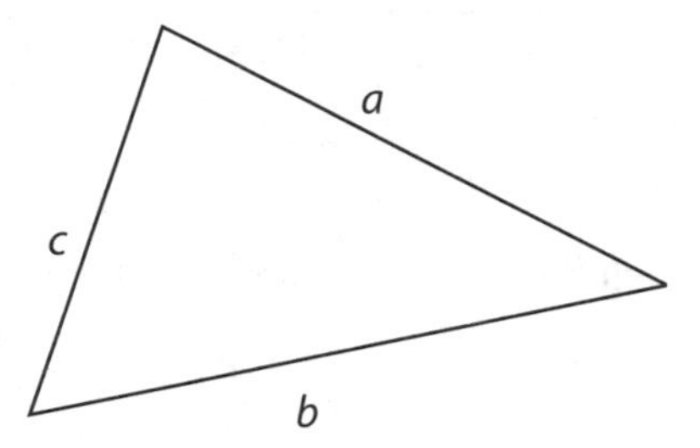

The perimeter of an *equilateral triangle* with sides each of length a is $P = 3a$.

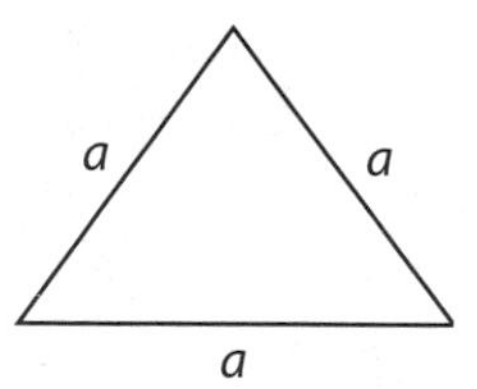

The perimeter of an *isosceles triangle* with congruent sides of length a and third side of length b is $P = 2a + b$.

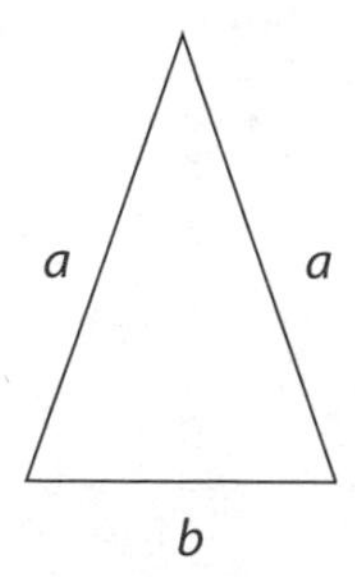

A *circle* is a unique figure. The perimeter of a circle is given the special name of *circumference*, *C*. It has no straight sides. The formula for C was discovered long ago after much thought and experimentation. It is simply $C = \pi d = 2\pi r$, where d is the length of the diameter of the circle and r is the length of its radius.

The usual approximation for π is 3.14, but if more accuracy is needed, you can use more decimal places. A calculator usually gives π to eight or ten decimal places.

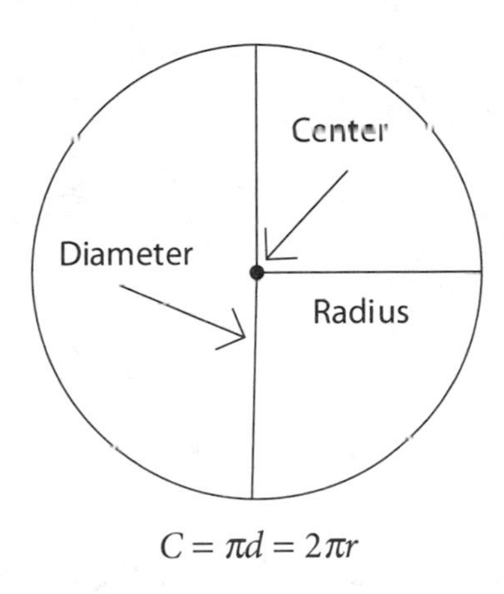

$$C = \pi d = 2\pi r$$

Problem Find the indicated perimeter or circumference.

a. Find the perimeter of a square that measures 3.5 cm on each side.

b. Find the perimeter of an isosceles triangle whose congruent sides are each 10 ft and the other side 6 ft.

c. Find the circumference of a circle whose radius is 20 cm. Use 3.14 to approximate π.

d. The circumference of the Earth is about 25,000 mi. What is the approximate diameter of the Earth? Use $\pi \approx 3.14$.

Solution

a. Find the perimeter of a square that measures 3.5 cm on each side.

Step 1. Sketch the figure.

Step 2. Use the formula for the perimeter of a square.

$P = 4s$

Step 3. Apply the formula to the figure and compute the perimeter.

$P = 4(3.5 \text{ cm}) = 14 \text{ cm}$

b. Find the perimeter of an isosceles triangle whose congruent sides are each 10 ft and the other side 6 ft.

Step 1. Sketch the figure.

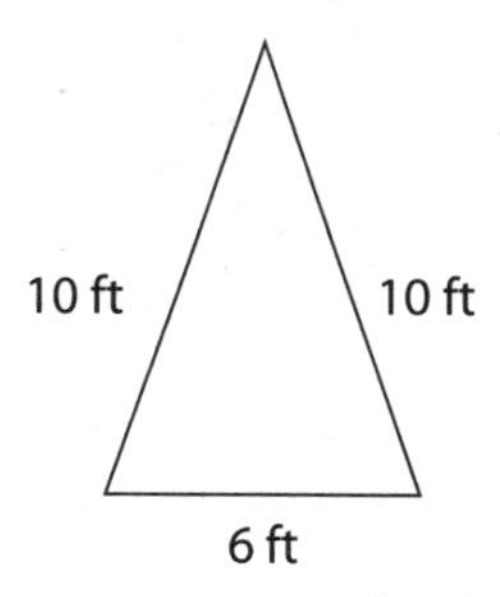

Step 2. Use the formula for the perimeter of an isosceles triangle with congruent sides of length a and third side of length b.

$P = 2a + b$

Step 3. Apply the formula to the figure and compute the perimeter.

$P = 2(10 \text{ ft}) + 6 \text{ ft} = 20 \text{ ft} + 6 \text{ ft} = 26 \text{ ft}$

c. Find the circumference of a circle whose radius is 20 cm. Use 3.14 to approximate π.

Step 1. Sketch the figure.

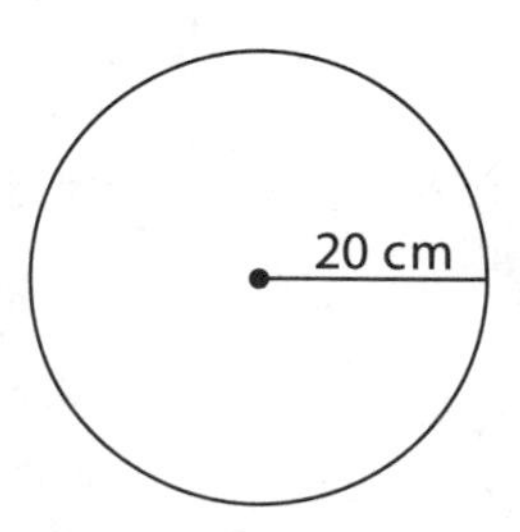

Step 2. Use the formula for the circumference of a circle.

$C = 2\pi r$

Step 3. Apply the formula to the figure and compute the circumference.

$$C = 2\pi(20 \text{ cm}) \approx 2(3.14)(20 \text{ cm}) = 125.6 \text{ cm}$$

The symbol "≈" means "is approximately equal to."

d. The circumference of the Earth is about 25,000 mi. What is the approximate diameter of the Earth? Use $\pi \approx 3.14$.

Step 1. Select the appropriate formula for the problem. The Earth is approximately a sphere, so the circumference is approximately that of a circle.

$$C = \pi d$$

Step 2. Apply the formula to the problem.

$$25{,}000 \text{ mi} = \pi d \approx (3.14)d$$

Step 3. Solve for d.

$$d = \frac{25{,}000 \text{ mi}}{3.14} \approx 7{,}962 \text{ mi}$$

Area

The *area* of a closed plane figure is a two-dimensional concept. It is the amount of surface enclosed by the boundary of the figure. For instance, in Figure 16.1, there are (2 in)(4 in) = 8(in)(in) = 8 in^2 of area enclosed by the rectangular figure shown.

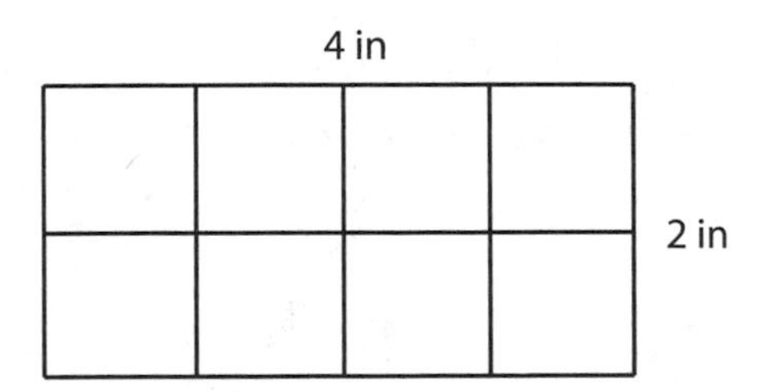

Each unit figure is a square of side length 1 in

Figure 16.1 Area of a plane figure

Area is always measured in square units such as square inches (in^2), square feet (ft^2), square miles (mi^2), and square meters (m^2). Regardless of the shape of the figure, the area units are always square units.

As with perimeter, special figures have special formulas for finding the area enclosed by the figure. The formula for the area of a rectangle is $A = lw$, where A is the area and l and w are the length and width of the sides of the rectangle.

$A = lw$

w

l

The area of a square is $A = s^2$.

$A = s^2$

s

The area of a triangle is $A = \frac{1}{2}bh$, where the *base*, b, can be any side of the triangle and the *height*, h, for that base is the perpendicular distance from the opposite vertex to that base (or an extension of it).

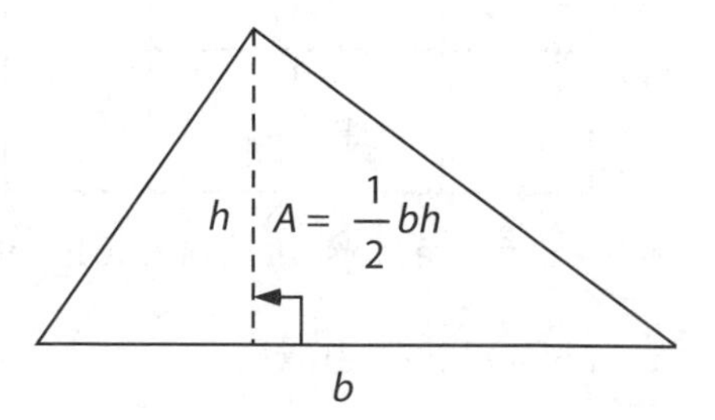

Like the circumference, the area of a circle involves the number π and is completely determined by the length of the radius. The formula for the area of a circle is $A = \pi r^2$, where r is the length of the radius.

The formula for the area of a circle is not $2\pi r^2$. This mistake is the result of confusing the formula for the area of a circle, $A = \pi r^2$, with $C = 2\pi r$, the radius form of the formula for the circumference. Each of these formulas has only one "2" in it. In the area formula, the "2" is an exponent indicating that the radius, r, is squared (because area is measured in square units). In the circumference formula, the "2" is a coefficient.

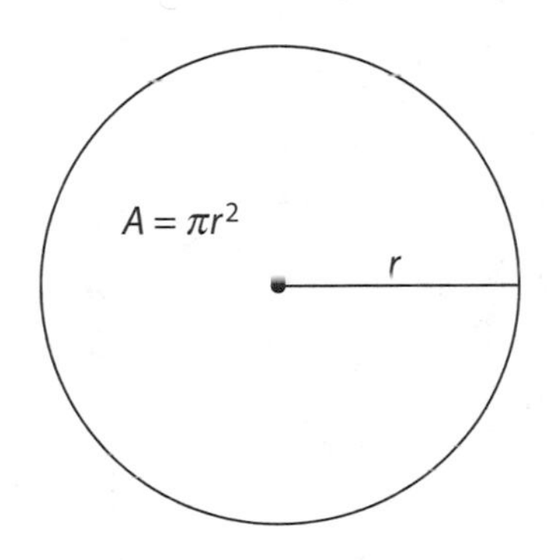

Problem Find the indicated area.

a. Find the area of a rectangle whose length is 6 in and width 4 in.

b. Find the area of a circle whose diameter is 24 cm.

c. Find the area of a triangle whose base measures 8 ft and height (to that base) measures 7 ft.

Solution

a. Find the area of a rectangle whose length is 6 in and width 4 in.

Step 1. Sketch the figure.

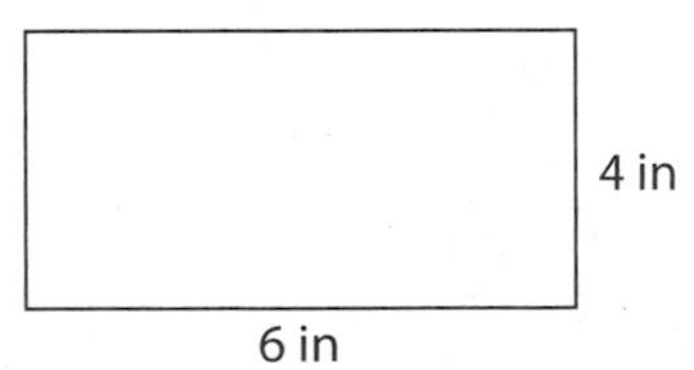

Step 2. Choose the appropriate area formula.

$A = lw$

Step 3. Apply the formula to the figure and compute the area.

$A = (6 \text{ in})(4 \text{ in}) = 24 \text{ in}^2$

b. Find the area of a circle whose diameter is 24 cm.

Step 1. Sketch the figure.

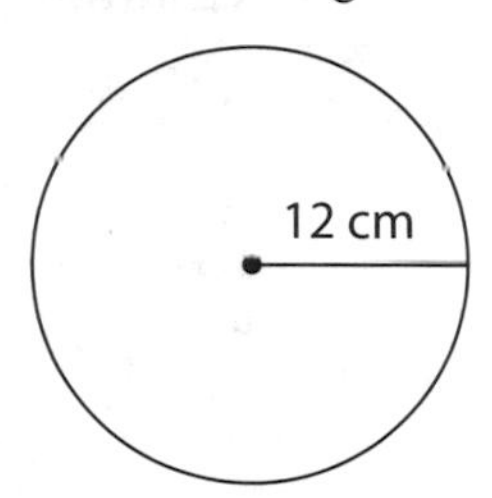

Note: The diameter is 24 cm, so the radius $= \frac{1}{2} \times 24\text{ cm} = 12\text{ cm}$.

Step 2. Choose the appropriate area formula.

$A = \pi r^2$

Step 3. Apply the formula to the figure and compute the area.

$A = \pi(12\text{ cm})^2 \approx (3.14)(144\text{ cm}^2) = 452.16\text{ cm}^2$

c. Find the area of a triangle whose base measures 8 ft and height (to that base) measures 7 ft.

Step 1. Sketch the figure.

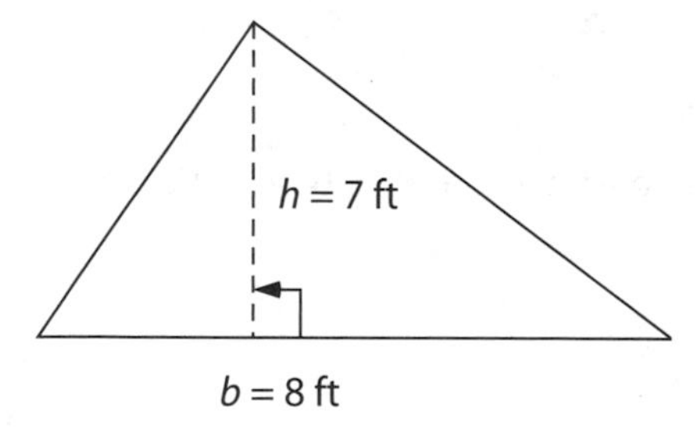

Step 2. Choose the appropriate area formula.

$$A = \frac{1}{2}bh$$

Step 3. Apply the formula to the figure and compute the area.

$$A = \frac{1}{2}(8\text{ ft})(7\text{ ft}) = 28\text{ ft}^2$$

Surface Area

The *surface area* of a solid three-dimensional figure is the area of the outside surface of the figure. As such, the surface area is a two-dimensional measure and has square units. If the figure is a rectangular prism (a box), as shown in Figure 16.2, then the surface area is the sum of the areas of all the faces.

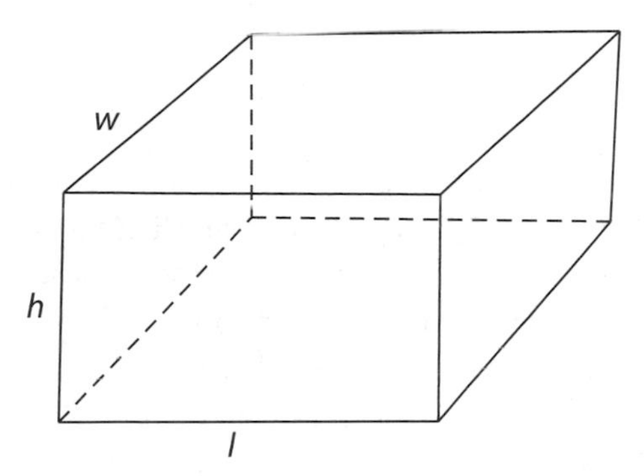

Figure 16.2 Rectangular prism

Problem Find the surface area of a box that has dimensions $l = 8$ in, $w = 6$ in, and $h = 5$ in.

Solution

Step 1. Sketch the figure.

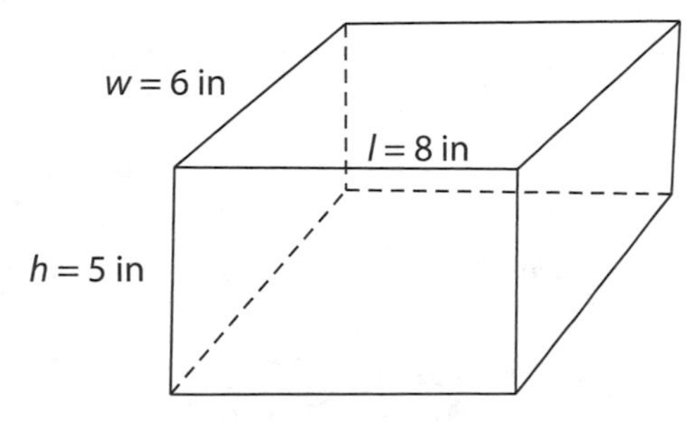

Step 2. Use the length and height to find the area of the front and rear faces.

Area of the front and rear faces = $2(8 \text{ in})(5 \text{ in}) = 80 \text{ in}^2$

Step 3. Use the height and width to find the area of the two side faces.

Area of the two side faces = $2(5 \text{ in})(6 \text{ in}) = 60 \text{ in}^2$

Step 4. Use the length and width to find the area of the top and bottom faces.

Area of the top and bottom faces = 2(6 in)(8 in) = 96 in^2

Step 5. Add the areas to get the final surface area.

Surface area = 80 in^2 + 60 in^2 + 96 in^2 = 236 in^2

Volume

Volume is a three-dimensional concept and is measured in *cubic* units. It is a measure of the space or capacity inside a three-dimensional closed figure such as a can, a cereal box, a room of a house, or a soccer ball. These types of measurements are very important to manufacturers of goods.

As with perimeter and area, special figures have special volume formulas. A rectangular prism (box) is a three-dimensional figure all of whose faces are rectangles. It has three characteristic measurements: length l, width w, and height h. The volume formula is $V = lwh$.

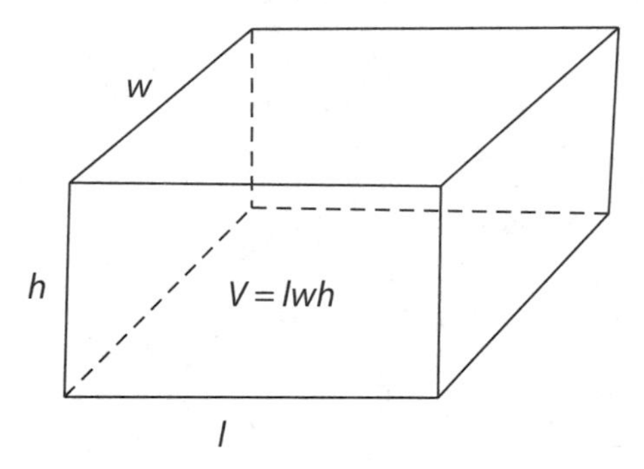

A *cylinder*, such as a can, has a circular base and top; consequently, the number π enters into the volume formula. The volume formula for a cylinder is $V = \pi r^2 h$, where r is the radius of the base circle and h is the height of the cylinder.

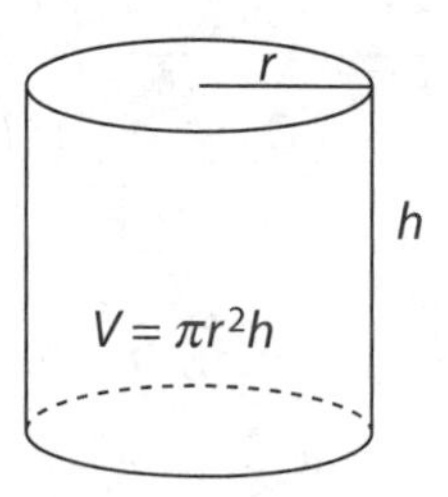

Notice that in both of the previous formulas, $V = (lw)h$ and $V = (\pi r^2)h$, the product in parentheses is the area of the base of the respective figures. If you let $B =$ the area of the base, then both formulas are the same, $V = Bh$. In fact, any prism or any cylinder has the same volume formula, $V = Bh$, where B is the area of the base and h is the height. This is the preferred format of these particular formulas in most of the current textbooks.

The following figures are triangular prisms, and the $V = Bh$ formula applies to them also. In this instance, the base areas are the areas of triangles. Notice that a prism does not have to rest on its "base."

In a cylinder or prism, the base and top are congruent figures in parallel planes, and they have the same area.

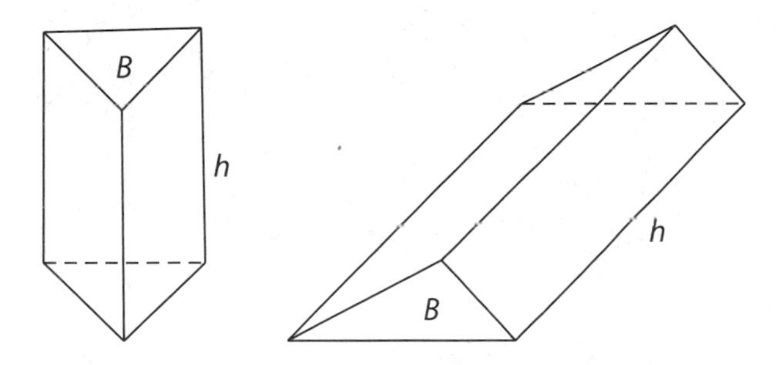

The volume, V, of a prism or cylinder that has a base of area B and height h is

$$V = Bh$$

A *sphere* has volume formula $V = \frac{4}{3}\pi r^3$, where r is the radius of the sphere. This formula can be derived with the tools of calculus.

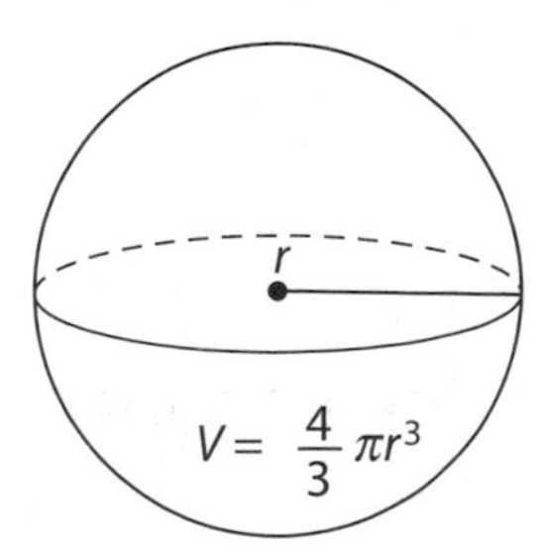

Problem Find the volume as indicated.

a. Find the volume of a sphere of radius 6 in.

b. The living room in a house has dimensions: $l = 30$ ft, $w = 20$ ft, and $h = 8$ ft. The owner of the house wants to know what volume of air the air conditioner will have to cool. This information will determine the size of air conditioner the owner will purchase. Find the volume of the living room.

c. Find the volume of a soda can that has a base radius of 2 in and a height of 5 in.

d. Find the volume of the triangular prism shown.

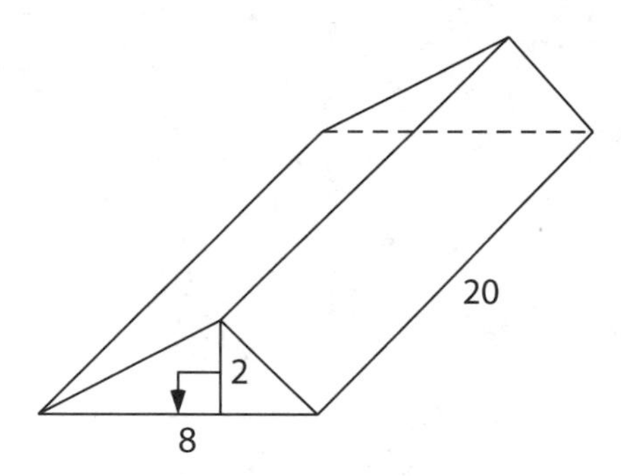

Solution

a. Find the volume of a sphere of radius 6 in.

Step 1. Sketch the figure.

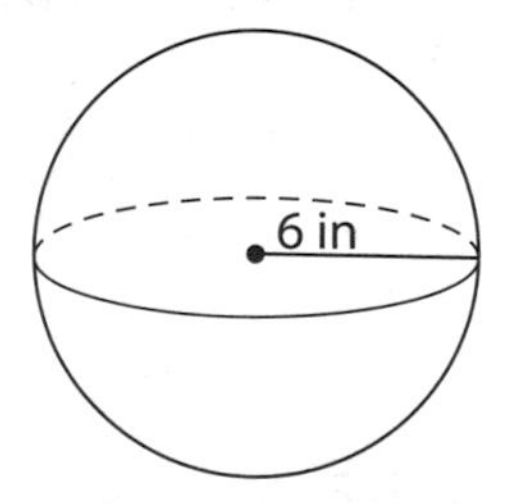

Step 2. Select the appropriate formula.

$$V = \frac{4}{3}\pi r^3$$

Step 3. Apply the formula to the problem and compute the volume.

$$V \approx \frac{4}{3}(3.14)(6 \text{ in})^3 \approx 904.32 \text{ in}^3$$

b. The living room in a house has dimensions: $l = 30$ ft, $w = 20$ ft, and $h = 8$ ft. The owner of the house wants to know what volume of air the air conditioner will have to cool. This information will determine the size of air conditioner the owner will purchase. Find the volume of the living room.

Step 1. Sketch the figure.

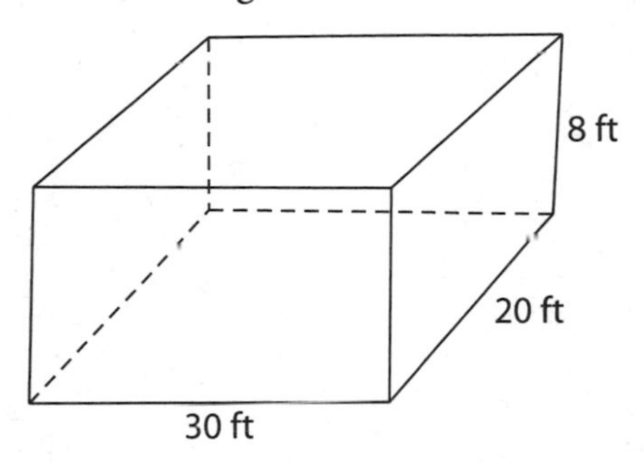

Step 2. Select the appropriate formula.

$V = Bh = (lw)h$

Step 3. Apply the formula to the problem and compute the volume.

$V = [(30 \text{ ft})(20 \text{ ft})]8 \text{ ft} = [600 \text{ ft}^2]8 \text{ ft} = 4{,}800 \text{ ft}^3$

c. Find the volume of a soda can that has a base radius of 2 in and a height of 5 in.

Step 1. Sketch the figure.

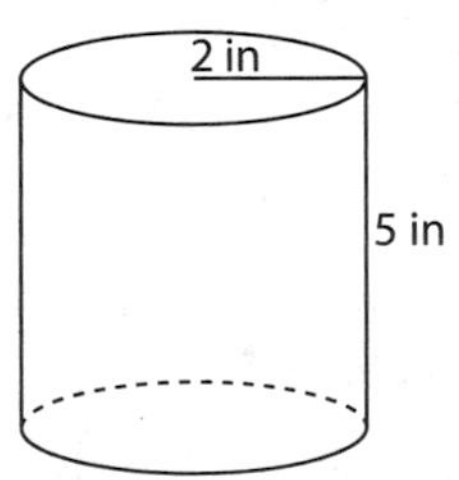

Step 2. Select the appropriate formula.

$V = Bh = (\pi r^2)h$

Step 3. Apply the formula to the problem and compute the volume.

$V = [\pi(2 \text{ in})^2]5 \text{ in} \approx [3.14(4 \text{ in}^2)]5 \text{ in} = 62.8 \text{ in}^3$

d. Find the volume of the triangular prism shown.

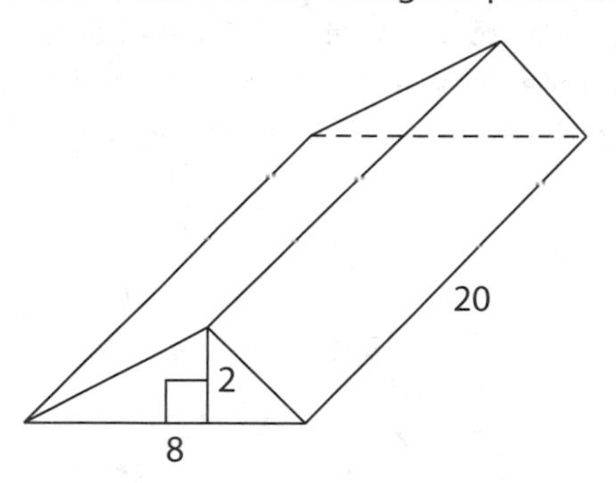

Step 1. Select the appropriate formula.

$V = Bh$

Step 2. Apply the formula to the problem and compute the volume.

$$V = Bh = \left(\frac{1}{2}(8)(2)\right)20 = 160 \text{ cubic units}$$

The exercises are designed to give you experience with different formulas and actually calculating perimeters, areas, and volumes. Some of these formulas you will remember from now on. Others will dim with time. The important thing is to remember there are such formulas and where to find them when needed. If you are comfortable with using them, you can put them to use to solve problems that might arise.

Exercise 16

1. The Mercedes-Benz Superdome in New Orleans is a hemisphere that has a diameter of 680 ft. What is the approximate circumference of the hemisphere?
2. How many 12 in by 12 in tiles are needed to cover a floor that is 20 ft by 30 ft?
3. A rectangular garden is 10 ft by 12 ft. A rectangular sidewalk 1 ft wide is built around the outside of the garden. What is the area of the complete garden region?
4. A rectangle and a square have equal areas. If the rectangle is 4 ft by 9 ft, how long is a side, *s*, of the square?

5. A lot is 21 ft by 30 ft. To support a fence, a post is needed at each corner and every 3 ft in between. How many posts are needed?
6. What is the area of the bottom surface of a circular swimming pool of diameter 20 ft?
7. What is the approximate volume of air trapped in the hemispherical top of the Mercedes-Benz Superdome in New Orleans? (See problem 1.)
8. A silo has a height of 100 ft and a base radius of 20 ft.
   a. How much grain (in ft^3) will the silo hold?
   b. If a bushel of wheat takes up 1 ft^3, how many bushels of wheat will the silo hold?
   c. If a bushel of wheat sells for $2, what is the value of a silo full of wheat?
9. Find the area of a triangle whose base measures 21 in and whose height measured to that base is 14 in.
10. Find the volume of the prism shown.

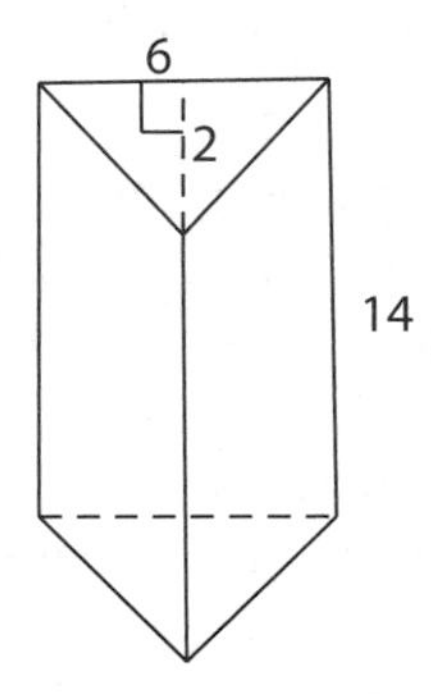

11. Find the surface area of a rectangular prism that has dimensions $l = 30$ ft, $w =$ 20 ft, and $h = 8$ ft.

17

Pythagorean Theorem

In this chapter, you learn about the Pythagorean theorem.

Right Triangle Concepts

A right triangle is a triangle with exactly one right (90°) angle. The longest side of the right triangle is opposite the 90° angle and is called the *hypotenuse*. The word *hypotenuse* derives from the Greek *hypo-* ("under") and *teinein* ("to stretch"). This interpretation was probably derived from the fact that the Pythagoreans, followers of the Greek mathematician Pythagoras, and other early geometers drew most of their triangles with the hypotenuse down ("under") as in Figure 17.1.

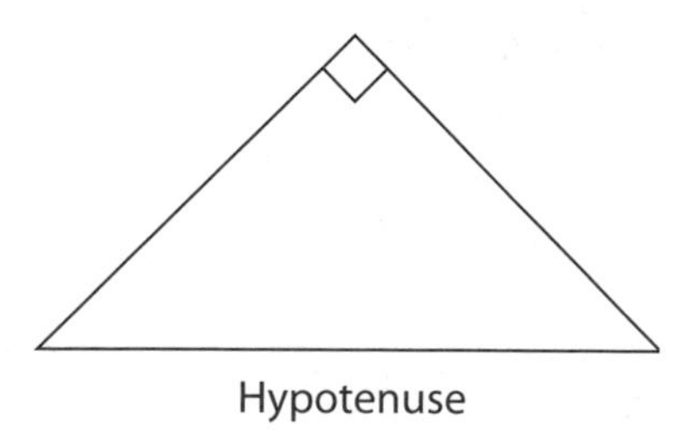

Figure 17.1 Right triangle

Pythagorean Theorem

The three sides of a right triangle have a special relationship, named after Pythagoras.

Pythagorean Theorem

If a right triangle has legs of lengths a and b and hypotenuse of length c, then $a^2 + b^2 = c^2$

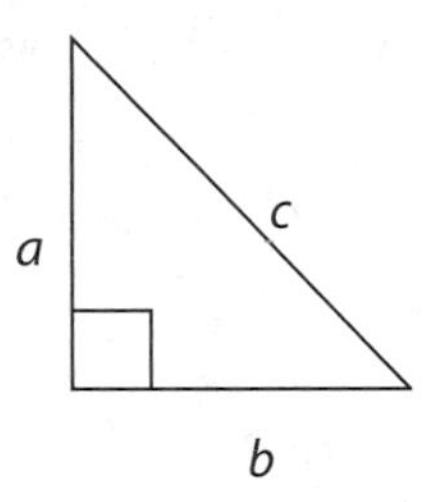

> The Pythagorean theorem is also written $c^2 = a^2 + b^2$. When using the theorem, choose whichever form is most convenient for the situation at hand.

This theorem is probably the most well-known theorem of mathematics and also one of the most used. Its discovery led not only to a great advance in geometry but also to the introduction of the completely unknown, at the time, set of irrational numbers.

The theorem was originally stated geometrically as the *square* on the hypotenuse is equal to the *sum of the squares* on the two legs, as depicted in Figure 17.2.

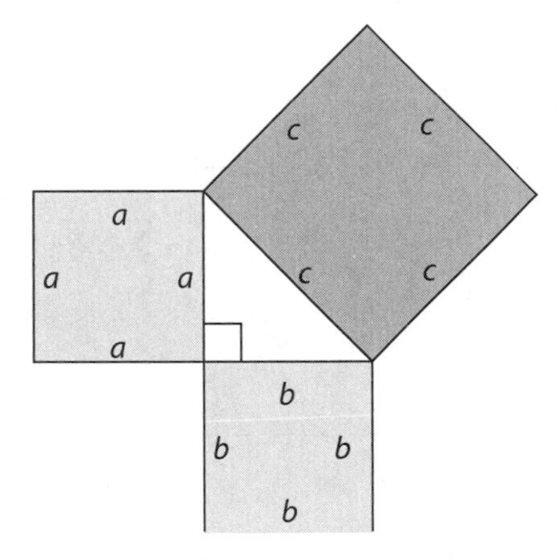

Figure 17.2 Depiction of the Pythagorean theorem

This formula is used so much today in algebra, calculus, physics, engineering, and many other disciplines that we tend to forget that it is a geometry theorem.

The example problems and the exercise problems are varied to illustrate the wide use of the theorem.

Problem Use the Pythagorean theorem to solve as indicated. Use your calculator and round answers, as needed.

a. Find the length of the diagonal of a rectangular garden that has dimensions of 30 ft by 40 ft.

b. Televisions are sized by the length of the diagonal of the face. Find the "size" of a portable TV whose rectangular face measures 27 in by 21 in.

c. A 25 ft ladder is placed against a wall with the base of the ladder 12 ft from the wall on level ground. How high up on the wall does the ladder reach?

d. Two people want to carry a tall mirror that is 9 by 9 ft square through a doorway that measures 3 ft by 8 ft. Will the mirror fit through the doorway?

e. Find the area of the triangle shown.

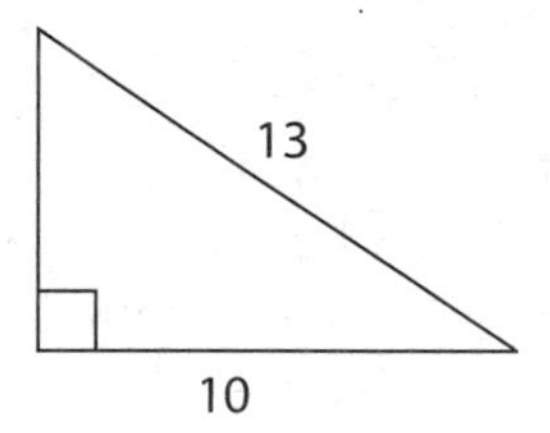

Solution

a. Find the length of the diagonal of a rectangular garden that has dimensions of 30 ft by 40 ft.

Step 1. Sketch the figure.

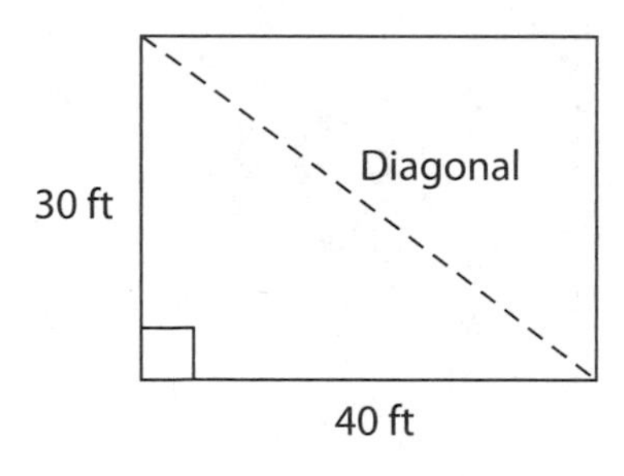

Step 2. From the figure, you can see that the diagonal is the hypotenuse of a right triangle having legs of 30 ft and 40 ft.

Step 3. Substitute the given values into the Pythagorean theorem, omitting the units for convenience.

$$c^2 = a^2 + b^2$$

$$c^2 = (30)^2 + (40)^2$$

$$c^2 = 900 + 1{,}600$$

$$c^2 = 2{,}500$$

Step 4. Solve for *c*.

To obtain *c*, you must find the square root of 2,500. Thus, you have

$c = \sqrt{2{,}500} = 50$

Step 5. State the answer.

The length of the diagonal of the garden is 50 ft.

b. Televisions are sized by the length of the diagonal of the face. Find the "size" of a portable TV whose rectangular face measures 27 in by 21 in.

Step 1. Sketch the figure.

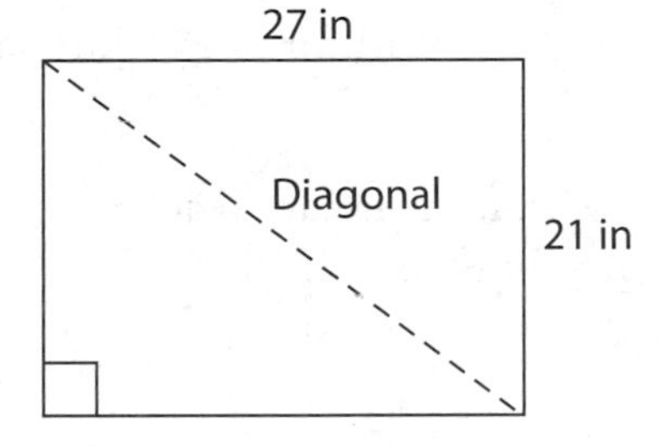

Step 2. From the figure, you can see that the diagonal is the hypotenuse of a right triangle having legs of 27 in and 21 in.

Step 3. Substitute these values into the Pythagorean theorem, omitting the units for convenience.

$c^2 = a^2 + b^2$

$c^2 = (27)^2 + (21)^2$

$c^2 = 729 + 441$

$c^2 = 1{,}170$

Step 4. Solve for *c*.

To obtain *c*, you must find the square root of 1,170. Thus, you have

$c = \sqrt{1{,}170} \approx 34.2.$

Step 5. State the answer.

The size of the TV is 34 in (rounded to the nearest whole number).

c. A 25 ft ladder is placed against a wall with the base of the ladder 12 ft from the wall on level ground. How high up on the wall does the ladder reach?

Step 1. Sketch the figure.

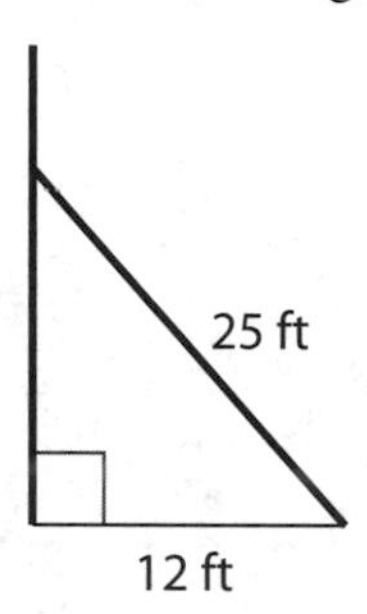

Step 2. The ladder, the ground, and the wall form a right triangle with the ladder as the hypotenuse.

Step 3. Let h = the distance up the wall the ladder reaches and substitute the numbers into the Pythagorean theorem.

$a^2 + b^2 = c^2$

$h^2 + 12^2 = 25^2$

$h^2 + 144 = 625$

$h^2 + 144 - 144 = 625 - 144$

$h^2 = 481$

Step 4. Solve for h.

To obtain h, you must find the square root of 481. Thus, you have

$h = \sqrt{481} \approx 21.93$

Step 5. Answer the question.

The ladder reaches approximately 22 ft up the wall.

d. Two people want to carry a tall mirror that is a 9 by 9 ft square through a doorway that measures 3 ft by 8 ft. Will the mirror fit through the doorway?

Step 1. Let m = length of the diagonal of the doorway. Sketch the figure.

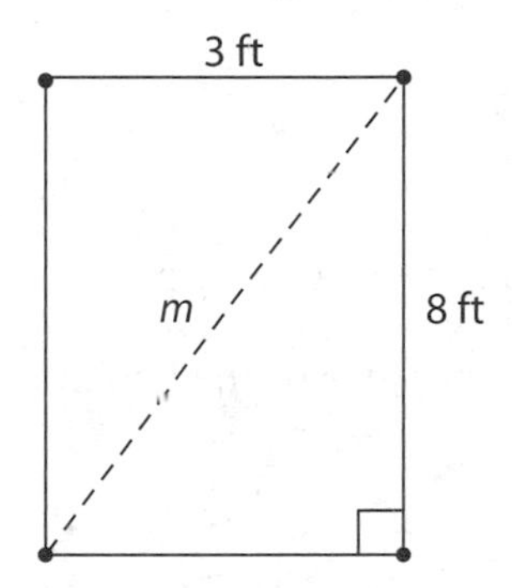

Step 2. Analyze the problem.

Because the doorway is only 8 ft high, the only way to get the mirror through the doorway is on a slant. Therefore, determine whether the length of the diagonal m of the doorway is greater than 9 ft, the length of an edge of the mirror.

Step 3. A right triangle is formed with m as the hypotenuse. Substitute the numbers into the Pythagorean theorem.

$c^2 = a^2 + b^2$

$m^2 = 3^2 + 8^2 = 9 + 64 = 73$

Step 4. Solve for m.

To obtain m, you must find the square root of 73. Thus, you have $m = \sqrt{73} \approx 8.54$.

Step 5. Answer the question.

The mirror will not fit through the doorway because the length of the diagonal of the doorway is about 8.54 ft, which is less than 9 ft, the length of an edge of the mirror.

e. Find the area of the triangle shown.

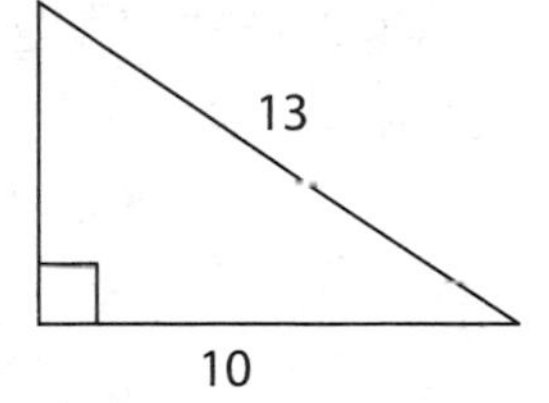

The height must be known in order to use the area formula for the triangle.

Step 1. Sketch the figure. Let h be the measure of the height.

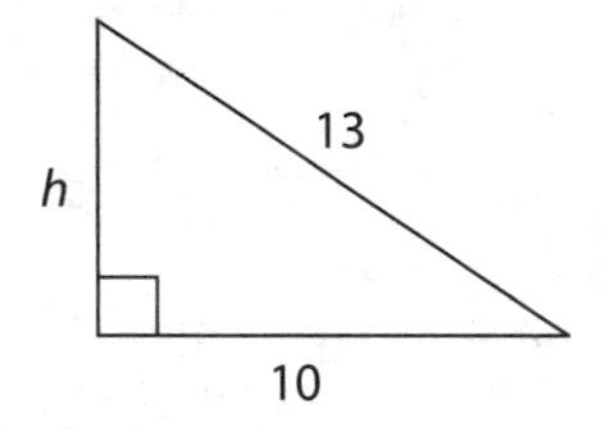

Step 2. Substitute into the Pythagorean theorem.

$$a^2 + b^2 = c^2$$

$$h^2 + 10^2 = 13^2$$

$$h^2 + 100 = 169$$

$$h^2 + 100 - 100 = 169 - 100$$

$$h^2 = 69$$

Step 3. Solve for h.

$h = \sqrt{69}$ (because 69 is not a perfect square, leave the answer in the radical form until the last calculation)

Step 4. Substitute the base and height into the area formula for the triangle.

$$A = \frac{1}{2}bh$$

$$A = \frac{1}{2}(10)\sqrt{69} = 5\sqrt{69}$$

Step 5. Answer the question.

The area of the triangle is $5\sqrt{69} \approx 41.53$ square units.

Converse of the Pythagorean Theorem

Triangles whose sides satisfy the Pythagorean theorem are right triangles. Thus, you have the following converse statement.

Converse of the Pythagorean Theorem

If a triangle has sides of lengths a, b, and c and $a^2 + b^2 = c^2$, then the triangle is a right triangle and c is the length of the hypotenuse.

Problem Is the triangle with sides of 8, 14, and 17 units a right triangle?

Solution

Step 1. Because 17 is the longest side, check whether the numbers satisfy the Pythagorean theorem.

$$17^2 \stackrel{?}{=} 8^2 + 14^2$$

$$289 \stackrel{?}{=} 64 + 196$$

$$289 \neq 260$$

Step 2. Answer the question.

The triangle is not a right triangle.

Exercise 17

In 1–4, solve for the unknown side. Use your calculator and round, as needed.

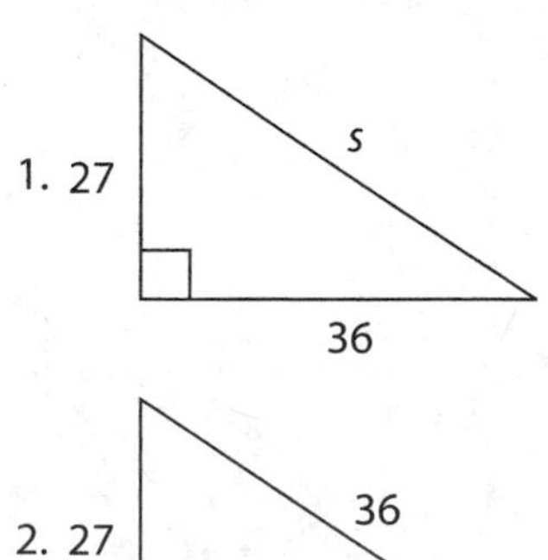

3.

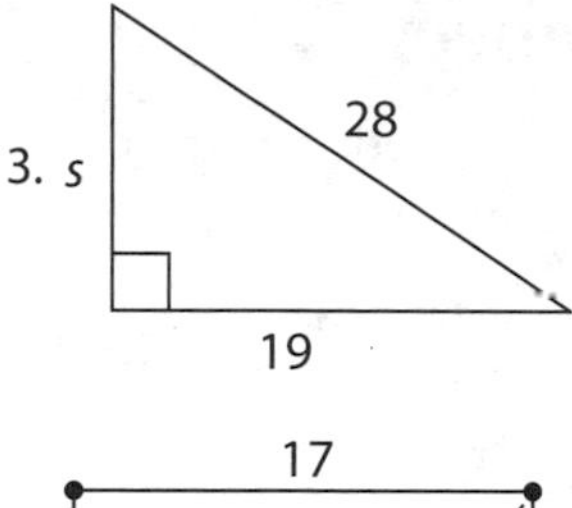

4.

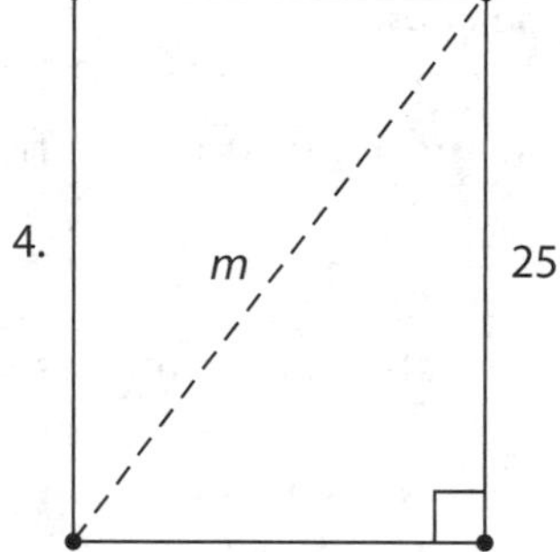

5. Is the triangle whose sides are 11, 15, and 19 units a right triangle?

6. Find the area of the region in the semicircle and outside the right triangle. (*Hint:* Recall the area formula for a circle.)

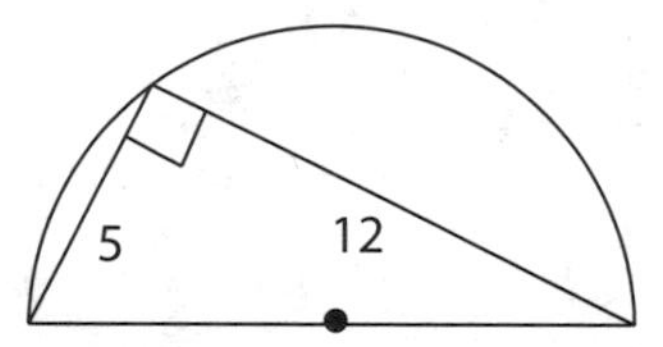

7. If a person in Texas travels 7 miles north, then 3 miles east, and finally 3 miles south, how far is the person from the starting point?

8. Find the perimeter of the square.

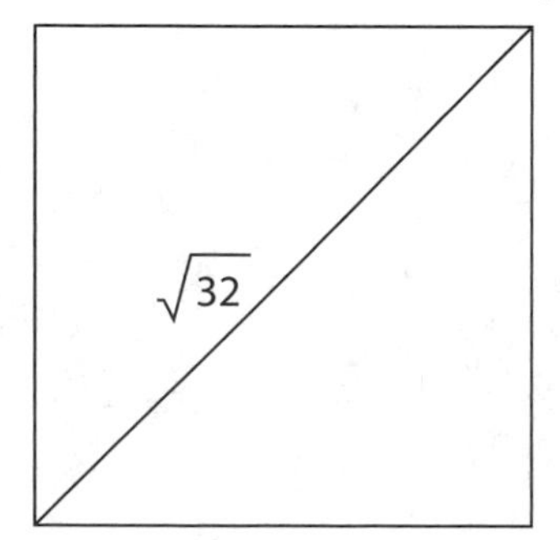

9. Is the triangle whose sides measure $\sqrt{6}$, $\sqrt{10}$, and 4 units a right triangle?

10. A ladder 4.0 m long is placed against a wall with the base of the ladder 0.8 m from the wall on level ground. How high up on the wall does the ladder reach?

18

Counting and Probability

In this chapter, you learn about counting and probability.

Counting Methods

You have different methods to find the number of ways to arrange or combine things. The most efficient method is to use the *counting principle*. The counting principle says that if you can do a first task in any one of m different ways, and after that you can do a second task in any one of n different ways, then you can do the first task followed by the second task in $m \times n$ different ways.

When you use the counting principle, *multiply*, don't add.

Problem A young boy is selecting a cap and a pair of tennis shoes to wear. He has a choice of a blue, red, orange, or yellow cap. He may select a black, brown, or white pair of tennis shoes. How many different ways can the boy choose one cap and one pair of tennis shoes?

Solution

Step 1. Determine how many tasks are involved.

Two tasks are involved: first, select a cap; next, select a pair of tennis shoes.

Step 2. Determine the number of ways to select a cap.

There are four colors to choose from, so the number of ways to select a cap is 4.

Step 3. Determine the number of ways to select a pair of tennis shoes.

There are three colors to choose from, so the number of ways to select a pair of tennis shoes is 3.

Step 4. Multiply the number of ways in step 2 by the number of ways in step 3.

(no. of ways to select a cap) $\times$ (no. of ways to select a pair of tennis shoes) $= 4 \times 3 = 12$

Note: "no." is an abbreviation for "number."

Step 5. State the answer.

There are 12 possible ways to select one cap and one pair of tennis shoes.

One popular method that you might have seen for working this problem is to list every possibility in a tree diagram. Here are the steps.

Step 1. List all the possibilities for the cap selection.

Blue Red Orange Yellow

Step 2. Below each cap selection, draw three "branches," one for each tennis shoe selection, and list the tennis shoe selections.

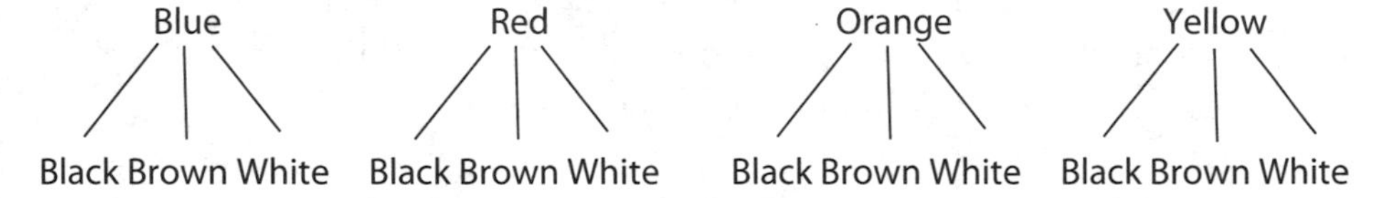

Step 3. Count the "leaves" at the base of the diagram to determine the number of possible ways to select one cap and one pair of tennis shoes.

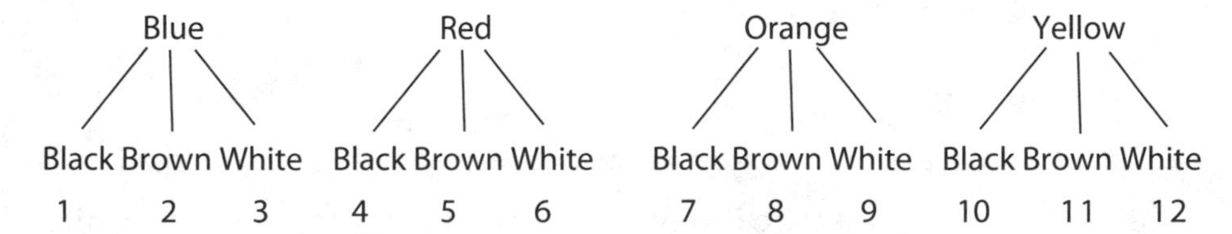

Step 4. State the answer.

There are 12 leaves, so there are 12 possible ways to select one cap and one pair of tennis shoes.

Notice in the tree diagram that each leaf represents a choice. For instance, the first leaf represents the choice of selecting a blue cap and black tennis shoes. This choice is different from the choice represented by the fourth leaf, which is selecting a red cap and black tennis shoes.

A third commonly used method is to list every possibility in an organized table. For the preceding problem, you proceed systematically in building the table, as shown here.

Step 1. List a blue cap color with each of the tennis shoe colors that can be selected.

CAP COLOR	TENNIS SHOE COLOR
Blue	Black
Blue	Brown
Blue	White

Step 2. List a red cap color with each of the tennis shoe colors that can be chosen.

CAP COLOR	TENNIS SHOE COLOR
Blue	Black
Blue	Brown
Blue	White
Red	Black
Red	Brown
Red	White

Step 3. List an orange cap color with each of the tennis shoe colors that can be chosen.

CAP COLOR	TENNIS SHOE COLOR
Blue	Black
Blue	Brown
Blue	White
Red	Black
Red	Brown
Red	White
Orange	Black
Orange	Brown
Orange	White

Step 4. List a yellow cap color with each of the tennis shoe colors that can be chosen.

CAP COLOR	***TENNIS SHOE COLOR***
Blue	Black
Blue	Brown
Blue	White
Red	Black
Red	Brown
Red	White
Orange	Black
Orange	Brown
Orange	White
Yellow	Black
Yellow	Brown
Yellow	White

Step 5. Count the number of rows in the table.

	CAP COLOR	***TENNIS SHOE COLOR***
1	Blue	Black
2	Blue	Brown
3	Blue	White
4	Red	Black
5	Red	Brown
6	Red	White
7	Orange	Black
8	Orange	Brown
9	Orange	White
10	Yellow	Black
11	Yellow	Brown
12	Yellow	White

There are 12 rows in the table.

Step 6. State the answer.

There are 12 possible ways to select one cap and one pair of tennis shoes.

Similar to the results of the tree diagram, each row in the table represents a choice. For instance, the 12th row represents the choice of selecting a yellow cap and white tennis shoes.

When you use a tree diagram or an organized table to count possibilities, be sure to proceed in a systematic manner, as illustrated in this chapter. Otherwise, you may overlook a possibility or count one more than once.

Tree diagrams and organized tables are useful counting methods when you have a limited number of choices, but they become increasingly unwieldy as the number of options increases. In contrast, the counting principle can be extended to accommodate any number of tasks, as shown in the following problem.

Problem How many different three-character codes consisting of one lowercase letter followed by two digits (0 to 9) are possible?

Solution

Step 1. Determine how many tasks are involved.

Three tasks are involved: first, select a lowercase letter; second, select a first digit; and third, select a second digit.

Step 2. Determine the number of ways to select a lowercase letter.

There are 26 lowercase letters, so the number of ways to select a lowercase letter is 26.

Step 3. Determine the number of ways to select the first digit.

There are 10 digits from which to choose, so the number of ways to select the first digit is 10.

Step 4. Determine the number of ways to select the second digit.

There are 10 digits from which to choose, so the number of ways to select the second digit is 10.

Step 5. Multiply the number of ways in step 2 by the number of ways in step 3 by the number of ways in step 4.

(no. of ways to select a letter) × (no. of ways to select a digit) × (no. of ways to select a digit) = $26 \times 10 \times 10 = 2{,}600$

Step 6. State the answer.

There are 2,600 possible three-digit codes consisting of one lowercase letter followed by two digits.

Basic Probability Concepts

Probability is a measure of the chance that an event will happen. In simple words, an event is something that happens subject to chance, for instance, heads showing face up when you flip a coin. You use the notation $P(E)$ as shorthand for "the probability that the event E will happen." If all possible outcomes are *equally likely*, then the probability that an event E will occur is determined this way:

In a probability problem, the number of possible outcomes is always greater than or equal to the number of outcomes favorable to the event. Always check to make sure that the denominator is *larger than or equal to* the numerator when you substitute values into the formula.

$$\text{Probability of event } E = P(E) = \frac{\text{number of outcomes favorable to } E}{\text{number of possible outcomes}}$$

"Equally likely" outcomes are outcomes that have the same chance of happening. "Favorable" outcomes are the outcomes that will result in the event E occurring. These are the outcomes you are looking for. Naturally, you can use other letters, such as A, B, and C, to represent events. It is customary to use uppercase letters for this purpose.

Do not use the formula $P(E) = \frac{\text{number of outcomes favorable to } E}{\text{number of possible outcomes}}$ unless all possible outcomes are equally likely. Doing so is a common error.

The following problem illustrates the important role that counting plays in computing probabilities.

Problem Find the indicated probability.

a. A fair coin is flipped one time, and the face-up side of the coin is observed. Find the probability that the up face is heads.

b. A bag contains five tiles, all identical in size and shape, that are numbered 10, 20, 30, 40, and 50. If a person picks out a single tile from the bag without looking, what is the probability that the number on the tile will be divisible by 20?

c. Suppose a three-character digital code for a lock box consists of one lowercase letter followed by two digits (0 to 9). Find the probability that a person can randomly guess the correct code for the lock box.

Solution

a. A fair coin is flipped one time, and the face-up side of the coin is observed. Find the probability that the up face is heads.

Step 1. Count the number of possible outcomes.

The possible outcomes of one flip of a coin are heads on the face up (H) or tails on the face up (T), so the number of possible outcomes is 2.

Step 2. Count the number of favorable outcomes.

There is one favorable outcome, H.

Step 3. Check whether all possible outcomes are equally likely.

The coin is a fair coin, so H and T are equally likely outcomes.

Step 4. Compute the probability.

$$P(\text{heads}) = \frac{\text{number of favorable outcomes}}{\text{number of possible outcomes}} = \frac{1}{2}$$

Step 5. State the answer.

On one flip of a fair coin, the probability of getting heads on the face-up side is $\frac{1}{2}$.

b. A bag contains five tiles, all identical in size and shape, that are numbered 10, 20, 30, 40, and 50. If a person picks out a single tile from the bag without looking, what is the probability that the number on the tile will be divisible by 20?

Step 1. Count the number of possible outcomes.

The possible outcomes are a 10, 20, 30, 40, or 50 on the tile, so there are five possible outcomes.

Step 2. Count the number of favorable outcomes.

There are two favorable outcomes for this event: drawing a tile numbered 20 or drawing a tile numbered 40.

Step 3. Check whether all possible outcomes are equally likely.

Given that the tiles are identical in size and shape and that the person draws a tile without looking, all possible outcomes are equally likely.

Step 4. Compute the probability.

$$P(20 \text{ or } 40) = \frac{\text{number of favorable outcomes}}{\text{number of possible outcomes}} = \frac{2}{5}$$

Step 5. State the answer.

The probability of drawing a tile whose number is divisible by 20 is $\frac{2}{5}$.

c. Suppose a three-character digital code for a lock box consists of one lowercase letter followed by two digits (0 to 9). Find the probability that a person can randomly guess the correct code for the lock box.

Step 1. Count the number of possible outcomes.

The number of possible codes (outcomes) = (no. of ways to select a letter)(no. of ways to select a digit)(no. of ways to select a digit) = $26 \times 10 \times 10 = 2{,}600$

Step 2. Count the number of favorable outcomes.

There is only one favorable outcome, the correct three-character digital code.

Step 3. Check whether all possible outcomes are equally likely.

Given that the person is trying to randomly guess the code, all possible outcomes are equally likely.

Step 4. Compute the probability.

$$P(\text{correct guess}) = \frac{\text{number of favorable outcomes}}{\text{number of possible outcomes}} = \frac{1}{2{,}600}$$

Step 5. State the answer.

The probability that a person can randomly guess the correct code for the lock box is $\frac{1}{2{,}600}$.

Probabilities can be expressed as fractions, decimals, or percents. In problem b earlier, the probability of drawing an even-numbered tile can be expressed as $\frac{2}{5}$, 0.4, or 40%.

The probability that an event is certain to happen is 1 or 100%. For instance, if you have a bag containing tiles numbered 10, 20, 30, 40, and 50 only, the probability of drawing a tile that has a multiple of 10 on it from the bag is 1 (because the numbers 10, 20, 30, 40, and 50 are all multiples of 10).

If an event cannot happen, then it has a probability of 0. For instance, the probability of drawing a tile with the number 60 on it from a bag containing tiles numbered 10, 20, 30, 40, and 50 only is 0 (because none of the tiles has 60 on it). Thus, the lowest probability is 0, and the highest probability is 1. All other probabilities fall between 0 and 1. Therefore, if you work a probability problem, and your answer is greater than 1 or your answer is negative, you've made an error! Go back and check your work.

Always remember to check whether the outcomes are equally likely before using the formula for probability. This situation is illustrated in the following problem.

Problem Suppose you spin a spinner that is one-fourth red, one-fourth yellow, and one-half green, as shown below. Find the probability that the spinner lands on red or green.

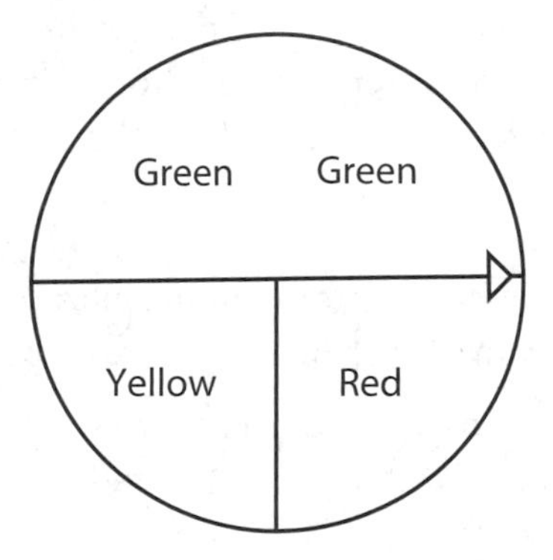

Solution

Step 1. Count the number of possible outcomes.

There are three possible outcomes because the spinner can land on red, yellow, or green.

Step 2. Count the number of favorable outcomes.

There are two favorable outcomes: landing on red or green.

Step 3. Check whether all possible outcomes are equally likely.

The three outcomes are not equally likely because the green section is larger than the other two sections.

Step 4. Compute the probability.

Given that the green and red sections occupy $\frac{1}{2}+\frac{1}{4}=\frac{3}{4}$ of the spinner, then $P(\text{red or green})=\frac{3}{4}\left(\text{not }\frac{2}{3}\right)$.

Step 5. State the answer.

The probability that the spinner lands on red or green is $\frac{3}{4}$.

Probability When Drawing with or Without Replacement

Sometimes a probability problem involves a two-step selection process. For instance, drawing a marble from a box of colored marbles, and then drawing a second marble from the same box. In this problem, it is important to consider whether the drawing is done "with replacement" or "without replacement." "With replacement" means that the first marble drawn is put back in the box before the second drawing takes place. "Without replacement" means that the first marble is not put back before the second drawing takes place. After accounting for "with replacement" or "without replacement," you multiply the probabilities of making each of the two selections, as illustrated in the following problem.

Problem Suppose a box contains 10 marbles: 3 red marbles, 5 blue marbles, and 2 green marbles, all identical except for color.

a. If you draw two marbles, one at a time, from the box without looking, what is the probability that you will draw two red marbles if your first draw is done *without* replacement?

b. If you draw two marbles, one at a time, from the box without looking, what is the probability that you will draw two red marbles if your first draw is done *with* replacement?

Solution

a. If you draw two marbles, one at a time, from the box without looking, what is the probability that you will draw two red marbles if your first draw is done *without* replacement?

Step 1. Find the probability of drawing a red marble on the first draw. Initially, there are 3 red marbles in the box of 10 marbles, so

$$P(\text{red on first draw})=\frac{\text{number of favorable outcomes}}{\text{number of possible outcomes}}=\frac{3}{10}$$

Step 2. Find the probability of drawing a red marble on the second draw. After a red marble is drawn and not put back, there are 2 red marbles in a box of 9 marbles, so *P*(red on second draw after drawing red on the first draw *without replacement*) $= \frac{\text{number of favorable outcomes}}{\text{number of possible outcomes}} = \frac{2}{9}$.

Step 3. Multiply the probability from step 1 by the probability from step 2.

$$P\text{ (2 red when draw } \textit{without} \text{ replacement)} = \frac{3}{10} \times \frac{2}{9} = \frac{6}{90} = \frac{1}{15}$$

Step 4. State the answer.

The probability of drawing two red marbles if the first draw is done *without* replacement is $\frac{1}{15}$.

b. If you draw two marbles, one at a time, from the box without looking, what is the probability that you will draw two red marbles if your first draw is done *with* replacement?

Step 1. Find the probability of drawing a red marble on the first draw.

Initially, there are 3 red marbles in the box of 10 marbles, so

$$P(\text{red on first draw}) = \frac{\text{number of favorable outcomes}}{\text{number of possible outcomes}} = \frac{3}{10}.$$

Step 2. Find the probability of drawing a red marble on the second draw.

After a red marble is drawn and put back, you again have 3 red marbles in a box of 10 marbles, so *P*(red on second draw after drawing red on the first draw *with replacement*)

$$= \frac{\text{number of favorable outcomes}}{\text{number of possible outcomes}} = \frac{3}{10}.$$

Step 3. Multiply the probability from step 1 by the probability from step 2.

$$P\text{ (2 red when draw } \textit{with} \text{ replacement)} = \frac{3}{10} \times \frac{3}{10} = \frac{9}{100}$$

Step 4. State the answer.

The probability of drawing two red marbles if the first draw is done *with* replacement is $\frac{9}{100}$.

This process of computing the probability of a sequence of events can be extended to any number of events. You should always consider whether "with replacement" or "without replacement" is a concern as you go from one event to the next.

Probability for Independent Events

Of course, when you are flipping coins or tossing numbered cubes, replacement is not a concern. Each flip of the coin or toss of the numbered cube is *independent* of the other flips or tosses. *Independent events* do not affect the probabilities of one another, so you simply multiply the probabilities in these cases. There are numerous other situations that involve independent events as well. In most cases, you can use your judgment to decide whether the events are independent.

> Coins (and other physical objects used in games of chance) do not have "memories," so do not make the mistake of thinking that a certain outcome is "due."

Problem Find the indicated probability.

a. A fair coin is flipped three times. Find the probability of getting three heads in a row.

b. A young boy wears a cap and a pair of tennis shoes to school every day. Suppose he randomly picks a blue, red, orange, or yellow cap, then randomly picks a black, brown, or white pair of tennis shoes. Find the probability that the boy will pick a red cap and black tennis shoes.

Solution

a. A fair coin is flipped three times. Find the probability of getting three heads in a row.

Step 1. Find the probability of obtaining heads on the first flip.

$$P(\text{heads on first flip}) = \frac{\text{number of favorable outcomes}}{\text{number of possible outcomes}} = \frac{1}{2}$$

Step 2. Find the probability of obtaining heads on the second flip.

$$P(\text{heads on second flip}) = \frac{\text{number of favorable outcomes}}{\text{number of possible outcomes}} = \frac{1}{2}$$

Step 3. Find the probability of obtaining heads on the third flip.

$$P(\text{heads on third flip}) = \frac{\text{number of favorable outcomes}}{\text{number of possible outcomes}} = \frac{1}{2}$$

Step 4. The events are independent, so multiply the probability from step 1 by the probability from step 2 by the probability from step 3.

$$P(3 \text{ heads in a row}) = \frac{1}{2} \times \frac{1}{2} \times \frac{1}{2} = \frac{1}{8}$$

Step 5. State the answer.

When a fair coin is flipped three times, the probability of getting three heads in a row is $\frac{1}{8}$.

b. A young boy wears a cap and a pair of tennis shoes to school every day. Suppose he randomly picks a blue, red, orange, or yellow cap, then randomly picks a black, brown, or white pair of tennis shoes. Find the probability that the boy will pick a red cap and black tennis shoes.

Step 1. Find the probability of picking a red cap.

$$P(\text{red cap}) = \frac{\text{number of favorable outcomes}}{\text{number of possible outcomes}} = \frac{1}{4}$$

Step 2. Find the probability of picking black tennis shoes.

$$P(\text{black tennis shoes}) = \frac{\text{number of favorable outcomes}}{\text{number of possible outcomes}} = \frac{1}{3}$$

Step 3. Because the boy is picking randomly and because the hat selection and shoe selection probabilities do not affect each other, the events are independent, so multiply the probability from step 1 by the probability from step 2.

$$P(\text{red cap and black tennis shoes}) = \frac{1}{4} \times \frac{1}{3} = \frac{1}{12}$$

Step 4. State the answer.

The probability the boy will pick a red cap and black tennis shoes is $\frac{1}{12}$.

Notice that the answer to problem b above is consistent with what you previously found; that is, that there are 12 possible ways for the boy to select one cap and one pair of tennis shoes. In only one of those 12 ways would the boy select a red cap and black tennis shoes. Thus, using the probability formula, you have

$$P(\text{red cap and black tennis}) = \frac{\text{number of favorable outcomes}}{\text{number of possible outcomes}} = \frac{1}{12}.$$

Exercise 18

1. Suppose you are playing a carnival game. You can pick door 1, door 2, or door 3, and then you select one of four curtains behind each door. How many different ways can you select one door and one curtain?
2. Suppose a code is dialed by means of three disks, each of which is stamped with 15 letters. How many three-letter codes are possible using the three disks?
3. A woman is ordering a sandwich with chips and a drink for lunch. She has a choice of three kinds of bread (white, whole wheat, or rye), four sandwich fillings (beef, chicken, ham, or turkey), two kinds of chips (potato chips or corn chips), and three kinds of drinks (cola, juice, or tea). How many different lunches can the woman order if she makes one bread selection, one sandwich filling selection, one chips selection, and one drink selection?
4. You toss a cube, whose faces are numbered 1, 2, 3, 4, 5, and 6, one time. What is the probability that a 5 appears face up?
5. You have a box of 50 marbles, all identical except for color. The box contains 20 blue, 10 red, 14 green, and 6 yellow marbles. If a person picks out a single marble from the box without looking, what is the probability that the marble will be green?
6. You have a box of 50 marbles, all identical except for color. The box contains 20 blue, 10 red, 14 green, and 6 yellow marbles. If a person picks out a single marble from the box without looking, what is the probability that the marble will be yellow or blue?
7. You have a box of 50 marbles, all identical except for color. The box contains 20 blue, 10 red, 14 green, and 6 yellow marbles. If a person picks out two marbles, one at a time, from the box without looking, what is the probability that the person will draw two red marbles if the first draw is done *without* replacement?
8. You have a box of 50 marbles, all identical except for color. The box contains 20 blue, 10 red, 14 green, and 6 yellow marbles. If a person picks out two marbles, one at a time, from the box without looking, what is the probability that the person will draw two red marbles if the first draw is done *with* replacement?
9. A fair coin is flipped five times. Find the probability of getting five heads in a row.
10. A spinner for a board game has three red sections, two yellow sections, four blue sections, and one green section. The sections are all of equal size. What is the probability of spinning red on the first spin and green on the second spin?

19

Mean, Median, Mode, and Range

In this chapter, you learn how to calculate the mean, median, mode, and range for a set of numbers. The mean, median, and mode are measures of central tendency. A *measure of central tendency* is a numerical value that describes a set of numbers (such as students' scores on a test) by attempting to provide a "central" or "typical" value of the numbers. The range is a *measure of variability*. It gives you an idea of the spread of a set of numbers.

Mean

The *mean* of a set of numbers is another name for the arithmetic average of the numbers in the set. To calculate the mean: first, sum the numbers; then, divide by how many numbers are in the set. Thus, you have the following formula:

$$\text{Mean} = \frac{\text{the sum of the numbers}}{\text{how many numbers in the set}}$$

Remember that the fraction bar indicates division.

Problem Find the mean for the given set of numbers.

a. {25, 43, 40, 60, 12}

b. {−7, 22, −7, 8, 16, 1}

c. {6.7, 7.6, 7.5, 6.9, 9.3, 6.7, 7.6, 8.5}

d. {0, 0, 0, 100, 100, 100}

e. {50, 50, 50, 50, 50, 50, 50}

f. {−70, −60, −50, −50, −40, −30}

The braces around a list of numbers mean that the numbers belong together as a set. You read {1,3,5} as "the set consisting of the numbers 1, 3, and 5."

Solution

a. {25, 43, 40, 60, 12}

Step 1. Sum the numbers.

$$25 + 43 + 40 + 60 + 12 = 180$$

When computing a mean, don't forget to divide by how many numbers you have.

Step 2. Divide by 5.

$$\frac{180}{5} = 36$$

If all the numbers are positive, then the mean of the numbers is also positive.

Step 3. State the answer.

The mean is 36.

b. {−7, 22, −7, 8, 16, 1}

Step 1. Sum the numbers.

$$-7 + 22 - 7 + 8 + 16 + 1 = 33$$

Step 2. Divide by 6.

$$\frac{33}{6} = 5.5$$

Step 3. State the answer.

The mean is 5.5.

c. {6.7, 7.6, 7.5, 6.9, 9.3, 6.7, 7.6, 8.5}

Step 1. Sum the numbers.

$$6.7 + 7.6 + 7.5 + 6.9 + 9.3 + 6.7 + 7.6 + 8.5 = 60.8$$

Step 2. Divide by 8.

$$\frac{60.8}{8} = 7.6$$

Step 3. State the answer.

The mean is 7.6.

d. {0, 0, 0, 100, 100, 100}

Step 1. Sum the numbers.

$$0 + 0 + 0 + 100 + 100 + 100 = 300$$

Step 2. Divide by 6.

$$\frac{300}{6} = 50$$

When you compute a mean, you divide by how many numbers you have, *including the number of zeros*, if any. Don't forget to count the zeros.

Step 3. State the answer.

The mean is 50.

e. {50, 50, 50, 50, 50, 50, 50}

Step 1. Sum the numbers.

$$50 + 50 + 50 + 50 + 50 + 50 + 50 = 350$$

Step 2. Divide by 7.

$$\frac{350}{7} = 50$$

As you might expect, if all the numbers are the same, the mean of the numbers is the common number.

Step 3. State the answer.

The mean is 50.

f. {–70,–60,–50,–50,–40,–30}

Step 1. Sum the numbers.

$$-70 - 60 - 50 - 50 - 40 - 30 = -300$$

Step 2. Divide by 6.

$$\frac{-300}{6} = -50$$

Step 3. State the answer.

The mean is –50.

If all the numbers are negative, then the mean of the numbers is also negative.

Weighted Average

A *weighted average* (or *weighted mean*) is an average computed by assigning weights to the numbers (e.g., test scores).

Problem A student scores 50, 40, and 96 on three exams. Find the weighted average of the student's three scores, where the score of 50 counts 20%, the score of 40 counts 30%, and the score of 96 counts 50%.

Solution

Step 1. Multiply each score by its corresponding "weight" and then add.

$$20\%(50) + 30\%(40) + 50\%(96) = 0.2(50) + 0.3(40) + 0.5(96) = 70$$

Step 2. Sum the weights.

$$20\% + 30\% + 50\% = 100\% = 1$$

Step 3. Divide the sum in step 1 by the sum in step 2.

$$\frac{70}{1} = 70$$

Step 4. State the answer.

The student's weighted average is 70.

Median

The *median* is the middle value or the arithmetic average of the two middle values in an *ordered* set of numbers. To find the median, do the following:

1. Put the numbers in order from least to greatest (or greatest to least).
2. Locate the middle number. If there is no single middle number, compute the arithmetic average of the two middle numbers.

Problem Find the median.

a. {25, 43, 40, 60, 12}

b. {−7, 22, −7, 8, 16, 1}

c. {6.7, 7.6, 7.5, 6.9, 9.3, 6.7, 7.6, 8.5}

d. {0, 0, 0, 100, 100, 100}

e. {50, 50, 50, 50, 50, 50, 50}

f. {−70,−60,−50,−50,−40,−30}

Solution

a. {25, 43, 40, 60, 12}

Step 1. Put the numbers in order from least to greatest.

12, 25, 40, 43, 60

Step 2. Locate the middle number.

40 is the middle number.

Step 3. State the answer.

The median is 40.

b. {−7, 22, −7, 8, 16, 1}

Step 1. Put the numbers in order from least to greatest.

−7, −7, 1, 8, 16, 22

Step 2. Locate the middle number.

There is no single middle number, so compute the arithmetic average of the two middle numbers, 1 and 8.

$$\frac{1+8}{2} - \frac{9}{2} = 4.5$$

Step 3. State the answer.

The median is 4.5.

c. {6.7, 7.6, 7.5, 6.9, 9.3, 6.7, 7.6, 8.5}

Step 1. Put the numbers in order from least to greatest.

6.7, 6.7, 6.9, 7.5, 7.6, 7.6, 8.5, 9.3

Step 2. Locate the middle number.

There is no single middle number, so compute the arithmetic average of the two middle numbers, 7.5 and 7.6.

$$\frac{7.5+7.6}{2} = \frac{15.1}{2} = 7.55$$

Step 3. State the answer.

The median is 7.55.

d. {0, 0, 0, 100, 100, 100}

Step 1. Put the numbers in order from least to greatest.

0, 0, 0, 100, 100, 100

Step 2. Locate the middle number.

There is no single middle number, so compute the arithmetic average of the two middle numbers, 0 and 100.

$$\frac{0+100}{2}=\frac{100}{2}=50$$

Step 3. State the answer.

The median is 50.

e. {50, 50, 50, 50, 50, 50, 50}

Step 1. Put the numbers in order from least to greatest.

50, 50, 50, 50, 50, 50, 50

Step 2. Locate the middle number.

The middle number is 50.

Step 3. State the answer.

The median is 50.

As with the mean, if all the numbers are the same, the median of the numbers is the common number.

f. {−70,−60,−50,−50,−40,−30}

Step 1. Put the numbers in order from least to greatest.

−70,−60,−50,−50,−40,−30

Step 2. Locate the middle number.

There is no single middle number, so compute the arithmetic average of the two middle numbers, −50 and −50.

$$\frac{-50+-50}{2}=\frac{-100}{2}=-50$$

If the two middle numbers are the same, then the arithmetic average is the common number.

Step 3. State the answer.

The median is −50.

As with the mean, if all the numbers are negative, then the median of the numbers is also negative.

Mode

In a set of numbers, the *mode* is the number (or numbers) that occurs most often (i.e., with the greatest frequency). A set of numbers in which each number occurs the same number of times has *no mode*. If only one number occurs most often, then the set is *unimodal*. If exactly two numbers occur with the same frequency that is more often than any of the other numbers, then the set is *bimodal*. If three or more numbers occur with the same frequency that is more often than any of the other numbers, then the set is *multimodal*.

Problem Find the mode, if any. For number sets that have modes, state whether the data set is unimodal, bimodal, or multimodal.

a. {25, 43, 40, 60, 12}

b. {–7, 22, –7, 8, 16, 1}

c. {6.7, 7.6, 7.5, 6.9, 9.3, 6.7, 7.6, 8.5}

d. {0, 0, 0, 100, 100, 100}

e. {50, 50, 50, 50, 50, 50, 50}

f. {–70,–60,–50,–50,–40,–30}

Solution

a. {25, 43, 40, 60, 12}

Step 1. For each number, determine how often it occurs.

Each number occurs only once.

Step 2. State the mode, if any.

There is no mode.

b. {–7, 22, –7, 8, 16, 1}

Step 1. For each number, determine how often it occurs.

The numbers 1, 8, 16, and 22 each occur only once.

The number –7 occurs twice.

Step 2. State the mode, if any.

The mode is –7, so the set is unimodal.

c. {6.7, 7.6, 7.5, 6.9, 9.3, 6.7, 7.6, 8.5}

Step 1. For each number, determine how often it occurs.

The numbers 6.9, 7.5, 8.5, and 9.3 each occur only once.

The numbers 6.7 and 7.6 each occur twice.

Step 2. State the mode, if any.

The modes are 6.7 and 7.6, so the set is bimodal.

d. {0, 0, 0, 100, 100, 100}

Step 1. For each number, determine how often it occurs.

The numbers 0 and 100 each occurs three times.

Step 2. State the mode, if any.

The modes are 0 and 100, so the set is bimodal.

e. {50, 50, 50, 50, 50, 50, 50}

Step 1. For each number, determine how often it occurs.

The number 50 occurs seven times.

Step 2. State the mode, if any.

The mode is 50, so the set is unimodal.

f. {–70,–60,–50,–50,–40,–30}

Step 1. For each number, determine how often it occurs.

The numbers –70, –60, –40, and–30 each occur only once.

The number –50 occurs twice.

Step 2. State the mode, if any.

The mode is –50, so the set is unimodal.

Range

The *range* describes the spread of a set of numbers. You find the range by computing the difference between the greatest number and the least number; that is,

Range = greatest number – least number

The range of a set of numbers is *always* nonnegative.

Problem Find the range.

a. {25, 43, 40, 60, 12}

b. {–7, 22, –7, 8, 16, 1}

c. {6.7, 7.6, 7.5, 6.9, 9.3, 6.7, 7.6, 8.5}

d. {0, 0, 0, 100, 100, 100}

e. {50, 50, 50, 50, 50, 50, 50}

f. {−70, −60,−50,−50,−40,−30}

Solution

a. {25, 43, 40, 60, 12}

Step 1. Identify the greatest number and the least number.

60 is the greatest number, and 12 is the least number.

Step 2. Compute the range.

range = greatest number − least number = 60 − 12 = 48

Step 3. State the answer.

The range is 48.

b. {−7, 22, −7, 8, 16, 1}

Step 1. Identify the greatest number and the least number.

22 is the greatest number, and −7 is the least number.

Step 2. Compute the range.

range = greatest number − least number = 22 − (−7) = 22 + 7 = 29

Step 3. State the answer.

The range is 29.

c. {6.7, 7.6, 7.5, 6.9, 9.3, 6.7, 7.6, 8.5}

Step 1. Identify the greatest number and the least number.

9.3 is the greatest number, and 6.7 is the least number.

Step 2. Compute the range.

range = greatest number − least number = 9.3 − 6.7 = 2.6

Step 3. State the answer.

The range is 2.6.

d. {0, 0, 0, 100, 100, 100}

Step 1. Identify the greatest number and the least number.

100 is the greatest number, and 0 is the least number.

Step 2. Compute the range.

range = greatest number − least number = 100 − 0 = 100

Step 3. State the answer.

The range is 100.

e. {50, 50, 50, 50, 50, 50, 50}

Step 1. Identify the greatest number and the least number.

50 is the greatest number, and 50 is the least number.

Step 2. Compute the range.

range = greatest number − least number = 50 − 50 = 0

Step 3. State the answer.

The range is 0.

> If all the numbers are the same, then the range of the numbers is 0.

f. {−70,−60,−50,−50,−40,−30}

Step 1. Identify the greatest number and the least number.

−30 is the greatest number, and −70 is the least number.

Step 2. Compute the range.

range = greatest number − least number = −30 − (−70) = −30 + 70 = 40

Step 3. State the answer.

The range is 40.

Exercise 19

For 1–4, find the mean.

1. {15, 33, 30, 50, 0}
2. {−4, 25, −4, 11, 19, 4}

3. {4.7, 5.6, 2.5, 4.9, 7.3, 4.7, 5.6, 6.5}
4. {−10, 0, 3, 3, 6, 16}
5. A student scores 80, 90, and 70 on three exams. Find the weighted average of the student's three scores, where the score of 80 counts 10%, the score of 90 counts 30%, and the score of 70 counts 60%.

For 6–10, find the median.

6. {15, 33, 30, 50, 0}
7. {−4, 25, −4, 11, 19, 4}
8. {4.7, 5.6, 2.5, 4.9, 7.3, 4.7, 5.6, 6.5}
9. {−10, 0, 3, 3, 6, 16}
10. {1, 1, 4, 4, 4, 10, 10, 10}

For 11–15, find the mode, if any. For number sets that have modes, state whether the set is unimodal, bimodal, or multimodal.

11. {15, 33, 30, 50, 0}
12. {−4, 25, −4, 11, 19, 4}
13. {4.7, 5.6, 2.5, 4.9, 7.3, 4.7, 5.6, 6.5}
14. {−10, 0, 3, 3, 6, 16}
15. {1, 1, 4, 4, 4, 10, 10, 10}

For 16–20, find the range.

16. {15, 33, 30, 50, 0}
17. {−4, 25, −4, 11, 19, 4}
18. {4.7, 5.6, 2.5, 4.9, 7.3, 4.7, 5.6, 6.5}
19. {−10, 0, 3, 3, 6, 16}
20. {150, 330, 300, 500, 0}

20

Graphical Representation of Data

This chapter presents reading and interpreting information from pictographs, bar graphs, circle graphs, line graphs, stem-and-leaf plots, histograms, and dot plots. Pictographs, bar graphs, and circle graphs display data organized into categories. Dot plots and stem-and-leaf plots show organized visual displays of data. Line graphs show a series of data points plotted over time. Histograms summarize data-using frequency or relative frequency of occurrence within intervals.

Pictographs

A *pictograph* represents data using one or more symbols (or images). A legend on the graph explains the meaning and the quantity each symbol represents. To read a pictograph, count the number of symbols in a row (or column) and multiply this number by the quantity the symbol represents as indicated by the legend.

Legends (or *Keys*) on graphs provide explanatory information to help with interpreting the graphic displays.

Sometimes, a fraction of a symbol is shown on a pictograph. In that case, approximate the fraction and use it accordingly.

Problem According to the graph shown, how many women responded "Yes" to the survey question posed?

Responses of 200 College-Age Women to the Survey Question "Do you own a car?"

Yes

No

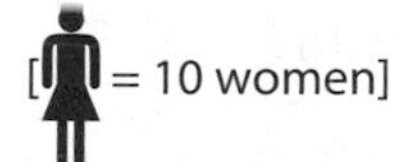

Solution

Step 1. Identify the type of graph.

The graph is a pictograph.

Step 2. Scan the graph to see what it is about.

The graph displays survey data about 200 college-age women's car ownership.

Step 3. Determine the meaning and the quantity represented by each symbol.

According to the graph's legend, each symbol represents 10 women.

Step 4. Count the number of symbols in the row labeled "Yes."

There are 15 symbols in the row labeled "Yes."

Step 5. Multiply the number in step 4 by the quantity represented by each symbol.

$15 \times 10 = 150$

Step 6. Answer the question.

According to the graph, 150 of the 200 women responded "Yes" to the survey question posed.

Bar Graphs

A *bar graph* uses rectangular bars to represent frequencies, percentages, or amounts. The bars correspond to different categories that are labeled at the base of the bars. The widths of the bars are equal. The length or height of a bar indicates the number, percentage, or amount for the category corresponding to that particular bar.

A scale (usually beginning with 0) marked at equally spaced intervals for measuring the height (or length) of the bars is shown on the graph. To read a bar graph, examine the scale to determine the units and the amount corresponding to each category. Then determine where the heights (or lengths) of the bars fall in relation to the category (or categories) of interest.

The bars in a bar graph may be arranged vertically or horizontally.

Problem According to the graph shown, what was the median annual earnings of all full-time, year-round workers ages 25–34 in 2015?

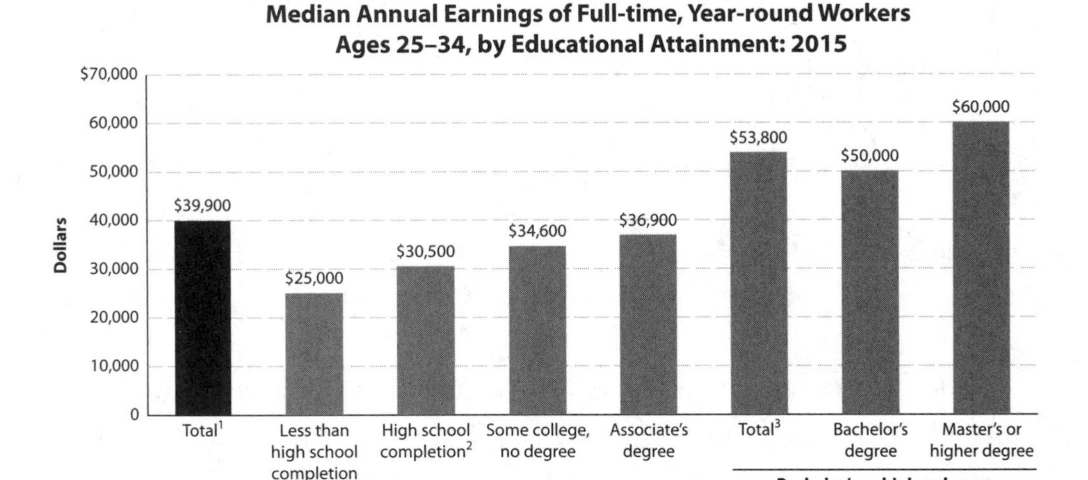

[1] Represents median annual earnings of all full-time, year-round workers ages 25–34.
[2] Includes equivalency credentials, such as the GED credential.
[3] Represents median annual earnings of full-time, year-round workers ages 25–34 with a bachelor's or higher degree.

NOTE: Data are based on sample surveys of the noninstitutionalized population, which excludes persons living in institutions (e.g., prisons or nursing facilities) and military barracks. *Full-time, year-round* workers are those who worked 35 or more hours per week for 50 or more weeks per year.

SOURCE: U.S. Department of Commerce, Census Bureau, Current Population Survey (CPS), "Annual Social and Economic Supplement," 2016. See *Digest of Education Statistics* 2016. table 502.30.

Solution

Step 1. Identify the type of graph.

The graph is a bar graph with vertically arranged bars.

Step 2. Scan over the graph to see what it is about.

The graph displays U.S. Department of Commerce, Census Bureau, Current Population Survey (CPS) data regarding the median annual earnings of full-time, year-round workers ages 25–34, by educational attainment, for the year 2015.

Step 3. Examine the graph's vertical scale.

The vertical scale is marked in increments of $10,000, from $0 to $70,000.

Step 4. Determine the height of the bar corresponding to the category of interest.

The height of the bar corresponding to full-time, year-round workers ages 25–34 was $39,900.

Step 5. Answer the question.

According to the graph, the median annual earnings of all full-time, year-round workers ages 25–34 was $39,900 in 2015.

Circle Graphs

A *circle graph* displays the relationship of distinct categories of data as portions of a whole, represented by a circle. The portions are labeled to show the categories for the graph. Percentages show the amount of the graph that corresponds to each category.

Problem According to the graph shown, how many students enrolled at University X are freshmen?

Distribution of 46,000 Students Enrolled at University X by Classification

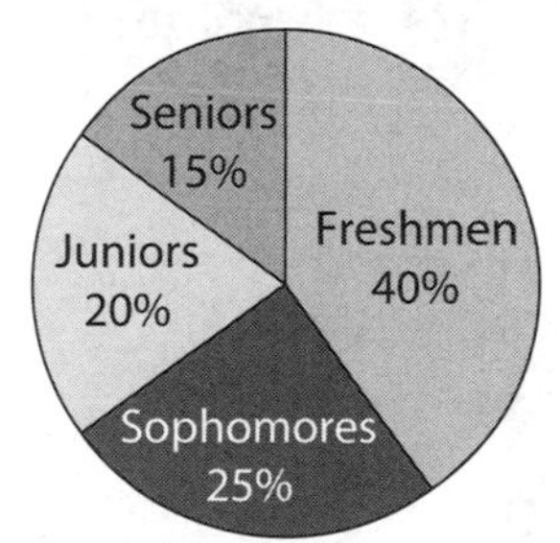

Solution

Step 1. Identify the type of graph.

The graph is a circle graph.

Step 2. Scan the graph to see what it is about.

The graph displays percentage data about freshmen, sophomores, juniors, and seniors enrolled at University X.

Step 3. Determine the percent of students enrolled as freshmen.
According to the graph, 40% of the students are freshmen.

Step 4. Multiply the percent in step 3 by the total number of students.
$40\% \times 46{,}000 = 0.4 \times 46{,}000 = 18{,}400$

Step 5. Answer the question.
According to the graph, 18,400 freshmen are enrolled at University X.

Line Graphs

A *line graph* shows a series of data points plotted over time. Line segments connect the plotted points. From left to right, an upward slant of a line segment indicates an increase, a downward slant indicates a decrease, and no slant indicates no change. To read the graph, determine the value corresponding to the time element of interest.

Sometimes, two or more line graphs are plotted on the same graph.

Problem According to the graph shown, what was the annual profit for the small online business in the year 2014?

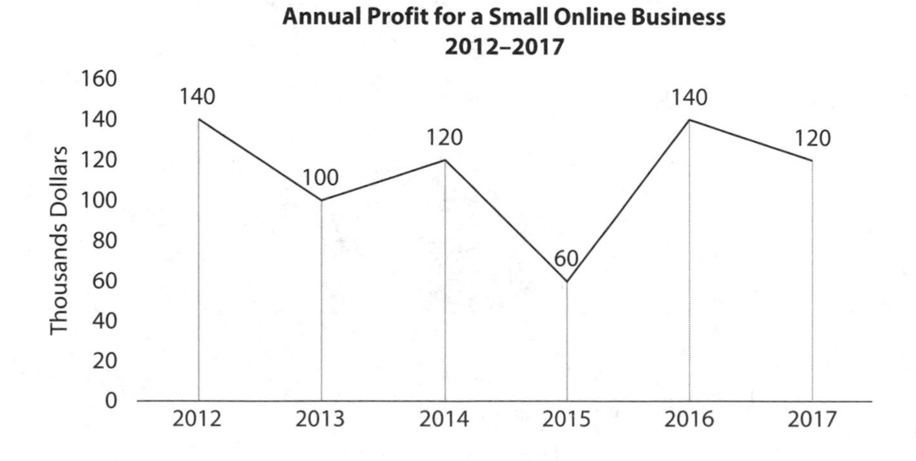

Solution

Step 1. Identify the type of graph.

The graph is a line graph.

Step 2. Scan the graph to see what it is about.

The graph displays information regarding the annual profit of a small online business from 2012 to 2017.

Step 3. Examine the graph's vertical scale.

The vertical scale is marked in increments of \$20,000, from \$0 to \$160,000.

Step 4. Determine the profit amount corresponding to the time of interest.

The annual profit amount corresponding to the year 2014 is \$120,000.

Step 5. Answer the question.

According to the graph, the annual profit for the online business was \$120,000 in the year 2014.

Stem-and-Leaf Plots

A *stem-and-leaf plot* is a visual display of data in which each data value is separated into two parts: a stem and a leaf. For a given data value, the leaf is the last digit, and the stem is the remaining digits. For example, for the data value 48, 4 is the stem and 8 is the leaf. Usually, the stems are listed vertically (from least to greatest), and the corresponding leaves for the data values are listed horizontally (from least to greatest) beside the appropriate stem.

When you create a stem-and-leaf plot, include a legend that explains what is represented by the stem and leaf so that the reader can interpret the information in the plot; for example, 4|8 = 48.

A useful feature of a stem-and-leaf plot is that the original data is retained and displayed in the plot.

Problem According to the graph shown, how many of the 25 guests who attended the fundraiser for the local PAWS dog shelter were 30 years old?

Ages of 25 Guests Who Attended a Fundraiser for a Local PAWS Dog Shelter

Stem	Leaf
2	0 1 4 5 9
3	0 0 2 3 4 8 9
4	2 3 5 7 8
5	1 3 8
6	5 6 8
7	2 8

Key: 4|8 = 48

Solution

Step 1. Identify the type of graph.

The graph is a stem-and-leaf plot.

Step 2. Scan the graph to see what it is about.

The graph shows the ages of guests who attended a fundraiser for a pet shelter.

Step 3. Count the number of ages that satisfy the question conditions.

There are two ages that are exactly 30.

Step 4. Answer the question.

According to the graph, 2 of the 25 guests who attended the fundraiser for the local PAWS dog shelter were 30 years old.

Histograms

A *histogram* groups data into a series of equal-length intervals ranging from the lowest to the highest data value. The heights of adjoining vertical bars above the intervals indicate the frequencies (or relative frequencies) of data values within intervals.

> The left and right endpoints for the intervals of a histogram are selected so that each data value clearly falls within one and only one interval.

Problem According to the graph, how many of the 100 ninth-grade boys are between 59.5 and 64.5 inches tall?

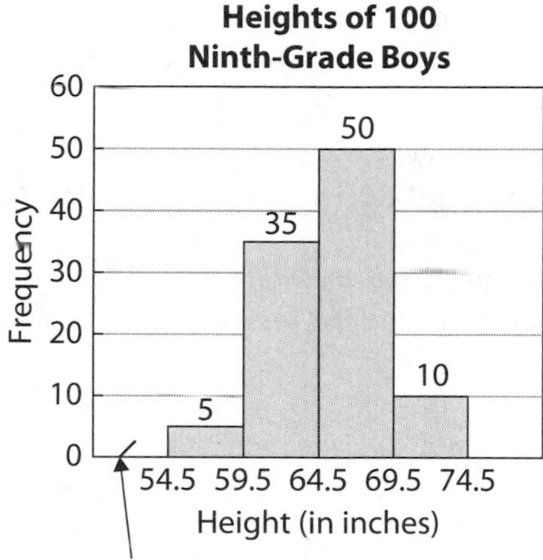

Solution

Step 1. Identify the type of graph.

The graph is a histogram.

Step 2. Scan the graph to see what it is about.

The graph summarizes the heights of 100 ninth-grade boys.

Step 3. Determine the number of heights that fall in the interval of interest.

There are 35 heights that fall between 59.5 and 64.5 inches.

Step 4. Answer the question.

According to the graph, 35 of the 100 ninth-grade boys are between 59.5 and 64.5 inches tall.

Dot Plots

A *dot plot* is a graph that shows the frequency of data values on a number line, and dots (or other similar symbols) are placed above each value to indicate the number of times that particular value occurs in the data set.

Problem According to the graph shown, how many of the 20 third graders read independently for exactly 15 minutes during free time in class?

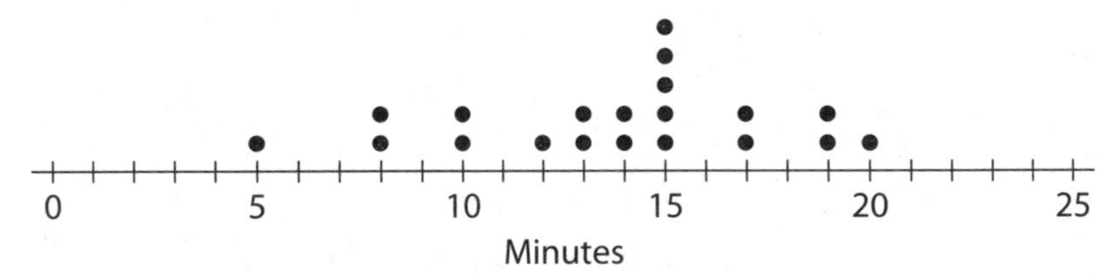

Key: Each dot represents one student.

Solution

Step 1. Identify the type of graph.

The graph is a dot plot.

Step 2. Scan the graph to see what it is about.

The graph shows the number of minutes that 20 third graders read independently during free time in class.

Step 3. Count the number of dots above the time of interest.

There are 5 dots above the time of 15 minutes.

Step 4. Answer the question.

According to the graph, 5 of the 20 third graders read independently for exactly 15 minutes during free time in class.

Exercise 20

1. According to the graph shown, how many women responded "No" to the survey question posed?

Responses of 200 College-Age Women to the Survey Question "Do you own a car?"

Yes

No

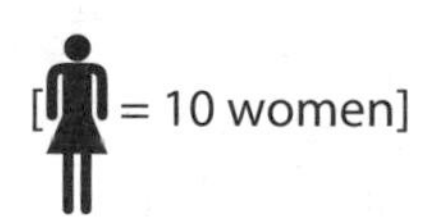

2. According to the graph shown, what is the median annual earnings of all full-time, year-round workers, ages 25–34, with a bachelor's or higher degree?

Median Annual Earnings of Full-time, Year-round Workers Ages 25–34, by Educational Attainment: 2015

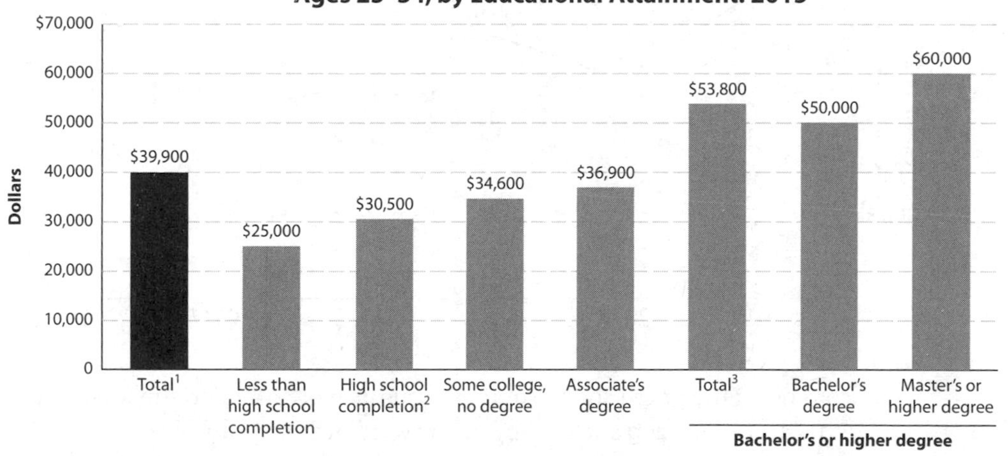

[1] Represents median annual earnings of all full-time, year-round workers ages 25–34.
[2] Includes equivalency credentials, such as the GED credential.
[3] Represents median annual earnings of full-time, year-round workers ages 25–34 with a bachelor's or higher degree.

NOTE: Data are based on sample surveys of the noninstitutionalized population, which excludes persons living in institutions (e.g., prisons or nursing facilities) and military barracks. *Full-time, year-round* workers are those who worked 35 or more hours per week for 50 or more weeks per year.

SOURCE: U.S. Department of Commerce, Census Bureau, Current Population Survey (CPS), "Annual Social and Economic Supplement," 2016. See *Digest of Education Statistics* 2016. table 502.30.

3. According to the graph shown, how many students enrolled at University X are freshmen or sophomores?

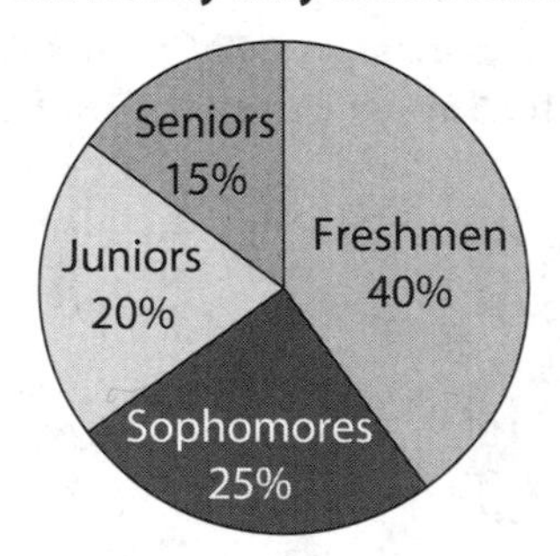

4. According to the graph shown, in which year did the small online business produce the least annual profit?

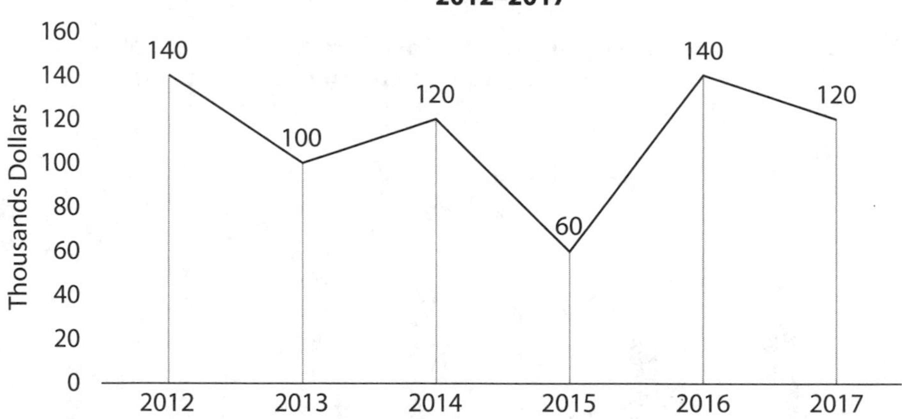

5. According to the graph shown, of the 25 guests who attended the fundraiser for the local PAWS dog shelter, how many were over 40 years old?

Ages of 25 Guests Who Attended a Fundraiser for a Local PAWS Dog Shelter

Stem	Leaf
2	0 1 4 5 9
3	0 0 2 3 4 8 9
4	2 3 5 7 8
5	1 3 8
6	5 6 8
7	2 8

Key: 4|8 = 48

6. According to the graph shown, how many of the 100 ninth-grade boys are at least 64.5 inches tall?

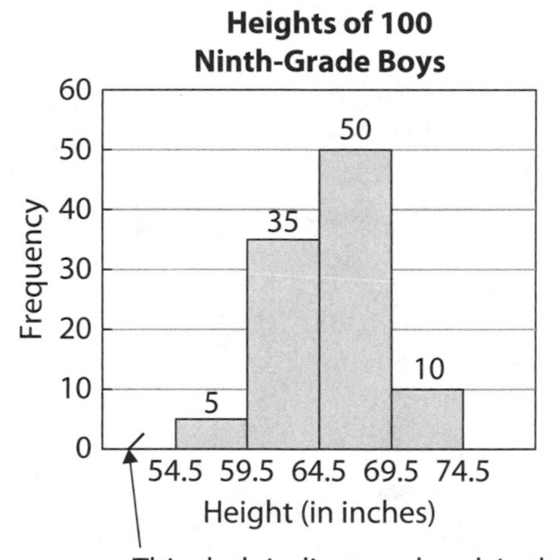

7. According to the graph shown, how many of the 20 third graders read independently for less than 15 minutes during free time in class?

Number of Minutes 20 Third Graders Read Independently During Free Time in Class

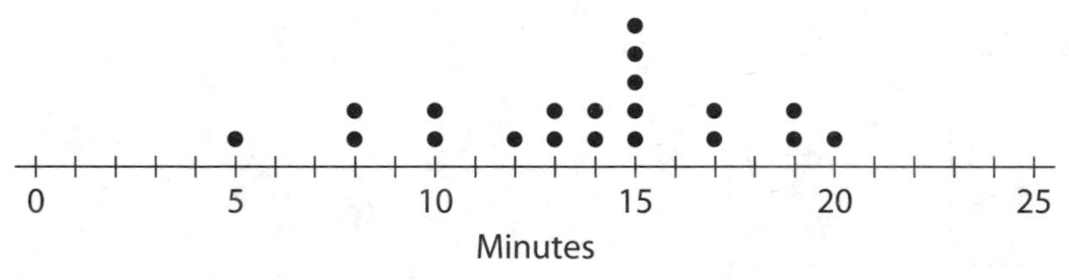

Key: Each dot represents one student.

Answer Key

Chapter 1 Numbers and Operations

Exercise 1

1. 10 is a natural number, a whole number, an integer, a rational number, and a real number.
2. -7.3 is a rational number and a real number.
3. -74 is an integer, a rational number, and a real number.
4. $-1,000$ is an integer, a rational number, and a real number.
5. $0.555\ldots$ is a rational number and a real number.
6. $-\frac{3}{4}$ is a rational number and a real number.
7. $\frac{345}{63}$ is a rational number and a real number.
8. 0 is a whole number, an integer, a rational number, and a real number.
9. $\sqrt{15}$ is an irrational number and a real number.
10. Undefined
11. 0
12. Undefined
13. Distributive property
14. Zero factor property
15. Associative property of multiplication

Chapter 2 Integers

Exercise 2

1. $|-45| = 45$

2. $|58| = 58$

3. $|-5| = 5$

4. "Negative nine plus the opposite of negative four equals negative nine plus four."

5. "Negative nine minus negative four equals negative nine plus four."

6. $-80 + -40 = -120$

7. $\frac{18}{-3} = -6$

8. $(-100)(-8) = 800$

9. $\frac{400}{2} = 200$

10. $-458 + 0 = -458$

11. $4(-3)(0)(999)(-500) = 0$

12. $\frac{0}{-56} = 0$

13. $\frac{-1{,}400}{-700} = 2$

14. $\frac{65}{-65} = -1$

15. $\frac{40}{0} =$ undefined

16. $(-3)(1)(-1)(-5)(-2)(2)(-10) = -600$

17. $(-3)(1)(-1)(-5)(-2)(0)(-10) = 0$

18.

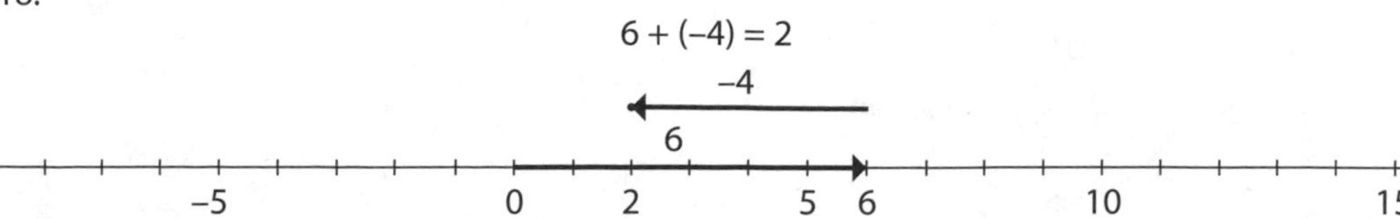

19.

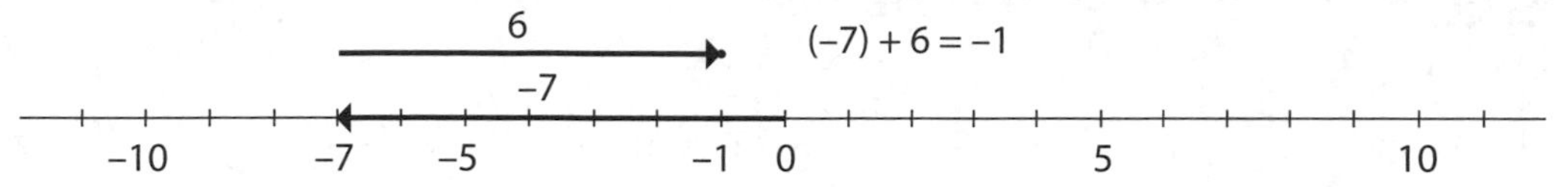

20.

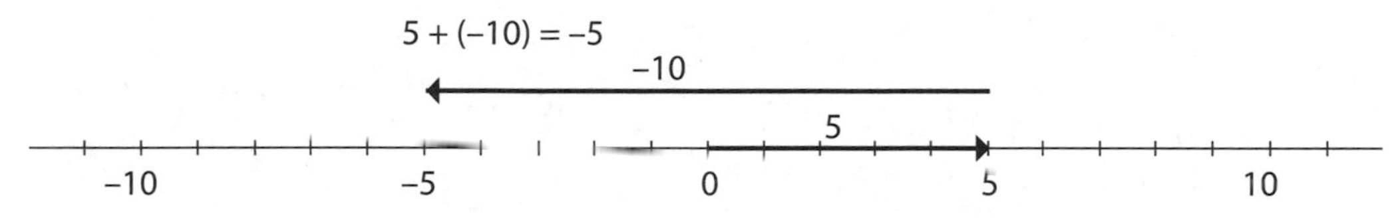

Chapter 3 Exponents

Exercise 3

1. "six to the fifth"
2. "negative five to the fourth"
3. "four to the zero"
4. "negative nine squared"
5. $(-4)(-4)(-4)(-4)(-4) = (-4)^5$
6. $8 \times 8 \times 8 \times 8 \times 8 \times 8 \times 8 = 8^7$
7. $2^8 = 2 \times 2 \times 2 \times 2 \times 2 \times 2 \times 2 \times 2 = 256$
8. $5^4 = 5 \times 5 \times 5 \times 5 = 625$
9. $(-4)^5 = (-4)(-4)(-4)(-4)(-4) = -1{,}024$
10. $0^9 = 0$
11. $(-2)^0 = 1$
12. 0^{-4} is the notation for $\frac{1}{0^4}$ or $\frac{1}{0}$, which is undefined.
13. $3^{-4} = \frac{1}{3^4} = \frac{1}{81}$
14. $(-15)^{-2} = \frac{1}{(-15)^2} = \frac{1}{225}$
15. $4^{-2} = \frac{1}{4^2} = \frac{1}{16}$

Chapter 4 Order of Operations

Exercise 4

1. $(5+7)6-10$
$=12\times 6-10$
$=72-10=62$

2. $(-7^2)(6-8)$
$=-49(-2)$
$=98$

3. $(2-3)(-20)$
$=(-1)(-20)$
$=20$

4. $3(-2)-\frac{10}{-5}$
$-6+\frac{10}{5}$
$=-6+2$
$=-4$

5. $9-\frac{20+22}{6}-2^3$
$=9-\frac{42}{6}-8$
$=9-7-8=-6$

6. $-2^2\cdot -3-(15-4)^2$
$=(-4)(-3)-(11)^2$
$=12-121=-109$

7. $5(11-3-6\cdot 2)^2$
$=5(11-3-12)^2$
$=5(-4)^2$
$=5(16)=80$

8. $-10-\frac{-8-(3\cdot -3+15)}{2}$
$=-10-\frac{-8-(-9+15)}{2}$
$=-10-\frac{-8-6}{2}$
$=-10-\frac{-14}{2}$
$=-10+7=-3$

9. $\frac{7^2-8\cdot 5+3^4}{3\cdot 2-36\div 12}$
$=\frac{49-40+81}{6-\frac{36}{12}}$
$=\frac{90}{6-3}=\frac{90}{3}=30$

10. $\frac{5-|-5|}{20^2}$
$=\frac{5-5}{400}$
$=\frac{0}{400}=0$

11. $\frac{3}{2}\left(-\frac{2}{3}\right)-\frac{1}{4}(-5)+\frac{15}{7}\left(-\frac{7}{3}\right)$
$=-1+\frac{5}{4}-5$
$=-1+1.25-5$
$=-4.75$

Chapter 5 Fractions

Exercise 5

1. $\frac{3}{8}+\frac{7}{16}=\frac{3\cdot 2}{8\cdot 2}+\frac{7}{16}$

$=\frac{6}{16}+\frac{7}{16}$

$=\frac{13}{16}$

2. $\frac{7}{11}-\frac{4}{7}=\frac{7\cdot 7-11\cdot 4}{11\cdot 7}$

$=\frac{49-44}{77}$

$=\frac{5}{77}$

3. $2\frac{3}{4}+\frac{11}{12}=\frac{11}{4}+\frac{11}{12}$

$=\frac{11\cdot 12+4\cdot 11}{4\cdot 12}$

$=\frac{132+44}{48}$

$=\frac{176}{48}=\frac{11}{3}\quad=3\frac{2}{3}$

4. $\frac{6}{5}\cdot\frac{15}{36}=\frac{90}{180}$

$=\frac{1}{2}$

5. $\frac{8}{9}\div\frac{4}{7}=\frac{8}{9}\cdot\frac{7}{4}$

$=\frac{56}{36}=\frac{14}{9}$

$=1\frac{5}{9}$

6. $\frac{3}{8}-\frac{7}{24}=\frac{3\cdot 3}{8\cdot 3}-\frac{7}{24}$

$=\frac{9}{24}-\frac{7}{24}$

$=\frac{2}{24}$

$=\frac{1}{12}$

7. $\left(\frac{6}{5}+\frac{3}{5}\right)\cdot\frac{5}{11}=\frac{9}{\cancel{5}}\cdot\frac{\cancel{5}}{11}$

$=\frac{9}{11}$

8. $\frac{7}{12}-\frac{4}{3}+\frac{5}{6}=\frac{7}{12}-\frac{4\cdot 4}{3\cdot 4}+\frac{5\cdot 2}{6\cdot 2}$

$=\frac{7}{12}-\frac{16}{12}+\frac{10}{12}$

$=\frac{7-16+10}{12}$

$=\frac{1}{12}$

9. $\frac{5}{11}\div\frac{5}{7}=\frac{5}{11}\cdot\frac{7}{5}$

$=\frac{\cancel{5}}{11}\cdot\frac{7}{\cancel{5}}$

$=\frac{7}{11}$

10. $\frac{7}{11}\div\frac{8}{11}=\frac{7}{\cancel{11}}\cdot\frac{\cancel{11}}{8}$

$=\frac{7}{8}$

11. $3\frac{3}{5}-2\frac{4}{7}=\frac{18}{5}-\frac{18}{7}$

$=18\left(\frac{1}{5}-\frac{1}{7}\right)$

$=18\left(\frac{1\cdot 7-5\cdot 1}{5\cdot 7}\right)$

$=18\left(\frac{2}{35}\right)=\frac{36}{35}$

$=1\frac{1}{35}$

12. $3\frac{2}{3}\div 4\frac{3}{5}=\frac{11}{3}\div\frac{23}{5}$

$=\frac{11}{3}\cdot\frac{5}{23}$

$=\frac{55}{69}$

Chapter 6 Decimals

Exercise 6

1.
$$\begin{array}{r} 45.716 \\ +3.920 \\ \hline 49.636 \end{array}$$

2.
$$\begin{array}{r} 23.3728 \\ -1.8264 \\ \hline 21.5464 \end{array}$$

3. Prediction: Five places

$$\begin{array}{r} 0.214 \\ \times 1.93 \\ \hline 0.41302 \end{array}$$ (using a calculator)

4. Prediction: Six places

$$\begin{array}{r} 1.21 \\ \times 0.0056 \\ \hline 0.006776 \end{array}$$ (using a calculator)

5. $0.014\overline{)0.1547}$ (move the decimal point)

$14\overline{)154.7}$

Answer: 11.05 (using a calculator)

6. $0.36\overline{)2.916}$ (move the decimal point)

   $36\overline{)291.6}$

   Answer: 8.1 (using a calculator)

7. Divide 2.917 by 0.37 and round to three places

   $0.37\overline{)2.917}$ (move the decimal point)

   $37\overline{)291.7}$

   Answer: 7.884 (using a calculator)

8. Multiply 6.678 by 0.37 and round to two places
   Answer: 2.47 (using a calculator)

9. Divide 3.977 by 0.0372 and round to three places

   $0.0372\overline{)3.977}$ (move the decimal point)

   $372\overline{)39770.}$

   Answer: 106.909 (using a calculator)

10. Multiply 45.67892 by 0.0374583 and round to four places
    Answer: 1.7111 (using a calculator)

Chapter 7 Percents

Exercise 7

1. $65\% = 0.\underset{\leftarrow}{65} = 0.65$
   (two places left)

2. $25.5\% = 0.\underset{\leftarrow}{25}5 = 0.255$
   (two places left)

3. $5\% = 0.\underset{\leftarrow}{05} = 0.05$
   (two places left)

4. $400\% = 4.\underset{\leftarrow}{00} = 4$
   (two places left)

5. $0.72 = 0\underset{\rightarrow}{72}.\% = 72\%$
   (two places right)

6. $0.325 = 0\underset{\rightarrow}{32}.5\% = 32.5\%$
(two places right)

7. $0.08 = 0\underset{\rightarrow}{08}.\% = 8\%$
(two places right)

8. $10 = 10\underset{\rightarrow}{00}.\% = 1{,}000\%$
(two places right)

9. $0.0075 = 0\underset{\rightarrow}{00}.75\% = 0.75\%$
(two places right)

10. $150\% = 150 \cdot \frac{1}{100} = \frac{150}{100} = \frac{150 \div 50}{100 \div 50} = \frac{3}{2} = 1\frac{1}{2}$

11. $\frac{1}{2}\% = \frac{1}{2} \cdot \frac{1}{100} = \frac{1}{200}$

12. $0.75\% = 0.75 \times \frac{1}{100} = \frac{0.75}{100} = \frac{75}{10{,}000} = \frac{75 \div 25}{10{,}000 \div 25} = \frac{3}{400}$

13. $66\frac{2}{3}\% = \frac{200}{3}\% = \frac{200}{3} \times \frac{1}{100} = \frac{200}{300} = \frac{200 \div 100}{300 \div 100} = \frac{2}{3}$

14. $\frac{3}{5} = 5\overline{)3.00}$ (quotient 0.60) $= 0.60 = 60\%$

15. $\frac{2}{3} = 3\overline{)2.00}$ (quotient 0.66 R2) $= 0.66\frac{2}{3} = 66\frac{2}{3}\%$

Chapter 8 Units of Measurement

Exercise 8

1. 8.25 km = __8,250__ m

   8.25×10^3 (three moves right) $= 8.25 \times 1{,}000 = 8{,}250$

2. 3 m = __300__ cm

   because 3×10^2 (three moves right) $= 3 \times 100 = 300$

3. 3 hr = __180__ min

   $\frac{3\ \cancel{hr}}{1} \times \frac{60\ min}{1\ \cancel{hr}} = \frac{180\ min}{1} =$ __180__ min

4. 7,200 sec = $\underline{\quad 2 \quad}$ hr

$$\frac{\overset{2}{\cancel{7,200}}\ \cancel{\text{sec}}}{1} \times \frac{1\text{ hr}}{\underset{1}{\cancel{3,600}}\ \cancel{\text{sec}}} = \frac{2\text{ hr}}{1} = \underline{\quad 2 \quad}\text{ hr}$$

5. 28 quarters = $\underline{\quad 70 \quad}$ dimes

$$\frac{28\ \cancel{\text{quarters}}}{1} \times \frac{25\text{ cents}}{\cancel{\text{quarters}}} = 700\text{ cents}$$

$$\frac{\overset{70}{\cancel{700}}\ \cancel{\text{cents}}}{1} \times \frac{1\text{ dime}}{\underset{1}{\cancel{10}}\ \cancel{\text{cents}}} = \frac{70\text{ dimes}}{1} = \underline{\quad 70 \quad}\text{ dimes}$$

6. 5 yd = $\underline{\quad 15 \quad}$ ft

$$\frac{5\ \cancel{\text{yd}}}{1} \times \frac{3\text{ ft}}{\cancel{\text{yd}}} = \underline{\quad 15 \quad}\text{ ft}$$

7. 720 min = $\underline{\quad 12 \quad}$ hr

$$\frac{\overset{12}{\cancel{720}}\ \cancel{\text{min}}}{1} \times \frac{1\text{ hr}}{\underset{1}{\cancel{60}}\ \cancel{\text{min}}} = \frac{12\text{ hr}}{1} = \underline{\quad 12 \quad}\text{ hr}$$

8. 3 yd^2 = $\underline{\quad 27 \quad}$ ft^2

$$\frac{3\ \cancel{\text{yd}^2}}{1} \times \frac{9\text{ ft}^2}{\cancel{\text{yd}^2}} = \frac{27\text{ ft}^2}{1} = \underline{\quad 27 \quad}\text{ ft}^2$$

9. 12 qt = $\underline{\quad 3 \quad}$ gal

$$\frac{\overset{3}{\cancel{12}}\ \cancel{\text{qt}}}{1} \times \frac{1\text{ gal}}{\underset{1}{\cancel{4}}\ \cancel{\text{qt}}} = \frac{3\text{ gal}}{1} = \underline{\quad 3 \quad}\text{ gal}$$

10. 10 in = $\underline{\quad 25.4 \quad}$ cm

$$\frac{10\ \cancel{\text{in}}}{1} \times \frac{2.54\text{ cm}}{1\ \cancel{\text{in}}} = \underline{\quad 25.4 \quad}\text{ cm}$$

11. 7500 kg = $\underline{\quad 7\,500\,000 \quad}$ g

$7{,}500 \times 10^3$ (three moves right) $= 7{,}500 \times 1{,}000 = 7{,}500{,}000$

12. 8.5 L = $\underline{\quad 8500 \quad}$ mL

8.5×10^3 (three moves right) $= 8.5 \times 1{,}000 = 8{,}500$

13. 15 quarters = _______ nickels

$$\frac{15\ \cancel{\text{quarters}}}{1} \times \frac{5\ \text{nickels}}{1\ \cancel{\text{quarters}}} = \frac{75\ \text{nickels}}{1} = \underline{\quad 75 \quad}\ \text{nickels}$$

14. $\frac{2}{3}$ yd = _______ ft

$$\frac{2\ \cancel{\text{yd}}}{\cancel{3}_1} \times \frac{\cancel{3}^1\ \text{ft}}{\cancel{\text{yd}}} = \frac{2\ \text{ft}}{1} = \underline{\quad 2 \quad}\ \text{ft}$$

15. Compare: $\frac{\$4.50}{8} = \$0.563$ to $\frac{\$4.99}{9} = \$0.554$.

The better buy is 9 oz for $4.99.

16. $250\ \text{m} \div 10^3$ (three moves left) $= 250\ \text{m} \div 1{,}000 = 0.25\ \text{km}$

17. $\frac{3\ \cancel{\text{lb}}}{1} \times \frac{16\ \text{oz}}{\cancel{\text{lb}}} = 48\ \text{oz}$

$48\ \text{oz} + 12\ \text{oz} = 60\ \text{oz}$

The package weighs 60 oz.

18. $\frac{\cancel{108}^{12}\ \cancel{\text{ft}^2}}{1} \times \frac{1\ \text{yd}^2}{\cancel{9}_1\ \cancel{\text{ft}^2}} = 12\ \text{yd}^2$

19. $\frac{3\ \cancel{\text{tbsp}}}{1} \times \frac{1\ \text{fl oz}}{2\ \cancel{\text{tbsp}}} = \frac{3\ \text{fl oz}}{2} = 1.5\ \text{fl oz}$

20. $\frac{5\ \cancel{\text{gal}}}{1} \times \frac{4\ \cancel{\text{qt}}}{1\ \cancel{\text{gal}}} \times \frac{2\ \cancel{\text{pt}}}{1\ \cancel{\text{qt}}} \times \frac{2\ \text{c}}{1\ \cancel{\text{pt}}} = 80\ \text{c}$

21. The thermometer is reading four tick marks above 10°. Since each interval between tick marks represents 1°, the thermometer is reading 4° above 10°, which is 14°C.

Chapter 9 Ratios and Proportions

Exercise 9

1. $\frac{14}{5}$ because both are in the same units

2. $n = \frac{50 \times 70}{250} = 14$

3. $\frac{2}{5} = \frac{x}{20 \text{ qt}}$

   $x = \frac{2 \times 20 \text{ qt}}{5} = 8 \text{ qt}$

4. $\frac{10 \text{ ft}}{8 \text{ ft}} = \frac{x}{60 \text{ ft}}$

   $x = \frac{10 \text{ ft} \times 60 \cancel{\text{ft}}}{8 \cancel{\text{ft}}} = 75 \text{ ft}$

5. $\frac{4 \text{ g}}{7 \text{ g}} = \frac{x}{56 \text{ g}}$

   $x = \frac{4 \text{ g} \times 56 \cancel{\text{g}}}{7 \cancel{\text{g}}} = 32 \text{ g}$

6. $\frac{\$12{,}000}{\$800} = \frac{x}{\$1{,}100}$

   $x = \frac{\$12{,}000(\cancel{\$}1{,}100)}{\cancel{\$}800} = \$16{,}500$

7. Let w = the number

   $\frac{92}{w} = \frac{80}{100}$

   $w = \frac{92 \times 100}{80} = 115$

8. Let p = the part

   $\frac{p}{80} = \frac{35}{100}$

   $p = \frac{80 \times 35}{100} = 28$

9. You will earn only one-half of a year's interest.
   Let A = the amount you will earn. Then $A = \frac{1}{2}(0.04)(5{,}000) = \$100$ is the amount of interest you will earn in 6 months.

10. $\frac{45}{120} = \frac{n}{100}$

    $n = \frac{45 \times 100}{120} = 37.5$

    $n\% = 37.5\%$

11. $\frac{4{,}500 \text{ km}}{300 \text{ min}} = \frac{x}{25 \text{ min}}$

    $x = \frac{25 \cancel{\text{min}}(4{,}500 \text{ km})}{300 \cancel{\text{min}}} = 375 \text{ km}$

12. $\frac{15 \text{ mi}}{1 \text{ in}} = \frac{x}{18 \text{ in}}$

    $x = \frac{18 \cancel{\text{in}} \times 15 \text{ mi}}{1 \cancel{\text{in}}} = 270 \text{ mi}$

13. $\frac{454.8}{7{,}580} = \frac{n}{100}$

    $n = \frac{454.8 \times 100}{7{,}580} = 6$

    $n\% = 6\%$

14. $\frac{36}{64} = \frac{5 \text{ g}}{x}$

    $x = \frac{64 \times 5 \text{ g}}{36} \approx 8.89 \text{ g}$

Chapter 10 Roots and Radicals

Exercise 10

1. $12 \times 12 = 144$ and $-12 \times -12 = 144$
   Thus, 12 and -12 are the two square roots of 144.

2. $\frac{5}{7} \times \frac{5}{7} = \frac{25}{49}$ and $-\frac{5}{7} \times -\frac{5}{7} = \frac{25}{49}$

   Thus, $\frac{5}{7}$ and $-\frac{5}{7}$ are the two square roots of $\frac{25}{49}$.

3. $0.8 \times 0.8 = 0.64$ and $-0.8 \times -0.8 = 0.64$
   Thus, 0.8 and -0.8 are the two square roots of 0.64.

4. $20 \times 20 = 400$ and $-20 \times -20 = 400$
   Thus, 20 and -20 are the two square roots of 400.

5. $6 \times 6 = 36$, so 6 is the positive square root of 36.
   Thus, $\sqrt{36} = 6$.

6. -9 is a negative number; square roots of negative numbers are not real numbers.
   Thus, $\sqrt{-9}$ is undefined for the set of real numbers.

7. $\frac{4}{5} \times \frac{4}{5} = \frac{16}{25}$, so $\frac{4}{5}$ is the positive square root of $\frac{16}{25}$.

   Thus, $\sqrt{\frac{16}{25}} = \frac{4}{5}$.

8. $\sqrt{25 + 144} = \sqrt{169}$

   $13 \times 13 = 169$, so 13 is the positive square root of 169.

   Thus, $\sqrt{25 + 144} = \sqrt{169} = 13$.

9. $-2 \times -2 \times -2 = -8$

   Thus, $\sqrt[3]{-8} = -2$.

10. $\frac{4}{5} \times \frac{4}{5} \times \frac{4}{5} = \frac{64}{125}$

    Thus, $\sqrt[3]{\frac{64}{125}} = \frac{4}{5}$.

Chapter 11 Algebraic Expressions

Exercise 11

1. The variable is s, and the constant is 4.
2. The numerical coefficient is -12.
3. The numerical coefficient is 1.
4. The numerical coefficient is $\frac{2}{3}$.
5. $-5x = -5(9) = -45$
6. $2xyz = 2(9)(-2)(-3) = 108$
7. $\frac{6(x+1)}{5\sqrt{x}-10} = \frac{6(9+1)}{5\sqrt{9}-10}$

   $= \frac{6 \cdot 10}{5(3)-10} = \frac{60}{5} = 12$

8. $\frac{-2|y|+5(2x-y)}{-6z+y^3} = \frac{-2|-2|+5(2\cdot 9-(-2))}{-6(-3)+(-2)^3}$

   $= \frac{-2(2)+5(18+2)}{18-8}$

   $= \frac{-4+100}{10} = \frac{96}{10} = \frac{48}{5} = 9\frac{3}{5}$

9. $x^2 - 8x - 9 = 9^2 - 8\cdot 9 - 9$

   $= 81 - 72 - 9 = 0$

10. $2y + x(y-z) = 2(-2) + 9((-2)-(-3))$

    $= -4 + 9(-2+3) = -4 + 9 = 5$

11. $(y+z)^{-3} = [(-2)+(-3)]^{-3}$

    $= (-5)^{-3}$

    $= \frac{1}{(-5)^3} = -\frac{1}{125}$

12. $5(x+6)$

    $= 5\cdot x + 5\cdot 6$

    $= 5x + 30$

13. $-3(x+4)$
$= -3 \cdot x + -3 \cdot 4$
$= -3x - 12$

14. $12 + (x^2 + y) = 12 + x^2 + y$

15. $8 - (2x - 4y) = 8 - 2x + 4y$

Chapter 12 Formulas

Exercise 12

1. $C = \frac{5}{9}(F - 32) = \frac{5}{9}(32 - 32) = 0$

2. $P = 4s = 4(10 \text{ m}) = 40 \text{ m}$

3. $c^2 = a^2 + b^2 = 7^2 + 24^2 = 49 + 576 = 625$
$c^2 = 625$
$c = \sqrt{625} = 25$

4. $A = 15 \text{ ft} \times 6 \text{ ft} = 90 \text{ ft}^2$

5. $I = \text{Prt} = (\$5{,}000)\left(\frac{3\%}{\text{yr}}\right)(10 \text{ yr}) = (\$5{,}000)\left(\frac{0.03}{\cancel{\text{yr}}}\right)(10 \cancel{\text{yr}}) = \$1{,}500$

6. $A = \pi r^2 \approx (3.14)(5 \text{ yd})^2 = (3.14)(25 \text{ yd}^2) = 78.5 \text{ yd}^2$

7. $V = \pi r^2 h = (3.14)(10 \text{ ft})^2(30 \text{ ft}) = (3.14)(100 \text{ ft}^2)(30 \text{ ft}) = 9{,}420 \text{ ft}^3$

8. $A = \frac{1}{2}bh = \frac{1}{2}(16)(20) = 160$

9. $V = \frac{1}{3}\pi r^2 h \approx \frac{1}{3}(3.14)(9 \text{ cm})^2(15 \text{ cm})$
$= \frac{1}{3}(3.14)(81 \text{ cm}^2)(15 \text{ cm}) = 1{,}271.7 \text{ cm}^3$

10. $F = \frac{9}{5}C + 32 = \frac{9}{\cancel{5}_1}\left(-\cancel{15}^{3}\right) + 32 = -27 + 32 = 5$

Chapter 13 Polynomials

Exercise 13

1. $x^2 - x + 1$ is a trinomial.
2. $125x^3 - 64y^3$ is a binomial.
3. $2x^2 + 7x - 4$ is a trinomial.
4. $-\frac{1}{3}x^5y^2$ is a monomial.
5. $2x^4 + 3x^3 - 7x^2 - x + 8$ is a polynomial.
6. $-15x + 17x = 2x$
7. $14xy^3 - 7x^3y^2$ is simplified.
8. $10x^2 - 2x^2 - 20x^2 = -12x^2$
9. $10 + 10x$ is simplified.
10. $2 + 4(x + 5) = 2 + 4x + 20$

    $= 4x + 22$
11. $12x^3 - 5x^2 + 10x - 60 + 3x^3 - 7x^2 - 1 = 15x^3 - 12x^2 + 10x - 61$
12. $(10x^2 - 5x + 3) + (6x^2 + 5x - 13) = 10x^2 - 5x + 3 + 6x^2 + 5x - 13$

    $= 10x^2 + 6x^2 - 5x + 5x + 3 - 13$

    $= 16x^2 - 10$
13. $(20x^3 - 3x^2 - 2x + 5) + (9x^3 + x^2 + 2x - 15) = 20x^3 - 3x^2 - 2x + 5 + 9x^3 + x^2 + 2x - 15$

    $= 20x^3 + 9x^3 - 3x^2 + x^2 - 2x + 2x + 5 - 15$

    $= 29x^3 - 2x^2 - 10$
14. $(10x^2 - 5x + 3) - (6x^2 + 5x - 13) = 10x^2 - 5x + 3 - 6x^2 - 5x + 13$

    $= 10x^2 - 6x^2 - 5x - 5x + 3 + 13$

    $= 4x^2 - 10x + 16$
15. $(20x^3 - 3x^2 - 2x + 5) - (9x^3 + x^2 + 2x - 15) = 20x^3 - 3x^2 - 2x + 5 - 9x^3 - x^2 - 2x + 15$

    $= 20x^3 - 9x^3 - 3x^2 - x^2 - 2x - 2x + 5 + 15$

    $= 11x^3 - 4x^2 - 4x + 20$

Chapter 14 Solving Equations

Exercise 14

1. $x-7=11$

   $x-7+7=11+7$

   $x=18$

2. $6x-3=13$

   $6x-3+3=13+3$

   $6x=16$

   $\frac{6x}{6}=\frac{16}{6}$

   $x=\frac{8}{3}=2\frac{2}{3}$

3. $x+3(x-2)=2x-4$

   $x+3x-6=2x-4$

   $4x-6=2x-4$

   $4x-6-2x=2x-4-2x$

   $2x-6=-4$

   $2x-6+6=-4+6$

   $2x=2$

   $\frac{2x}{2}=\frac{2}{2}$

   $x=1$

4. $\frac{x+3}{5}=\frac{x-1}{2}$

   $\frac{10}{1}\cdot\frac{(x+3)}{5}=\frac{10}{1}\cdot\frac{(x-1)}{2}$

   $2(x+3)=5(x-1)$

   $2x+6=5x-5$

   $2x+6-5x=5x-5-5x$

   $-3x+6=-5$

   $-3x+6-6=-5-6$

   $-3x=-11$

   $\frac{-3x}{-3}=\frac{-11}{-3}$

   $x=\frac{11}{3}$

5. Two more than three times a number x = four less than six times the number x.

   $3x+2=6x-4$

   $3x+2-6x=6x-4-6x$

   $-3x+2=-4$

   $-3x+2-2=-4-2$

   $-3x=-6$

   $\frac{-3x}{-3}=\frac{-6}{-3}$

   $x=2$

6. $R=35\%=0.35$, $B=500$, $P=?$

   $P=RB$

   $P=(0.35)(500)$

   $P=175$

   175 is 35% of 500.

7. $R=60\%=0.6$, $B=?$, $P=90$

   $P=RB$

   $90=(0.6)B$

   $0.6B=90$

   $\frac{0.6B}{0.6}=\frac{90}{0.6}$

   $B=150$

   90 is 60% of 150.

8. $R=?$, $B=\$144$, $P=\$21.60$

   $P=RB$

   $\$21.60=R(\$144)$

   $\$144R=\$21.60$

   $\frac{\$144R}{\$144}=\frac{\$21.60}{\$144}$

   $R=0.15=15\%$

   \$21.60 is 15% of \$144.

9. $R = ?, B = 80, P = 70$

   $P = RB$

   $70 = R(80)$

   $80R = 70$

   $\frac{80R}{80} = \frac{70}{80}$

   $R = 0.875$

   $R = 0.875 = 87.5\%$

   The student's percent grade is 87.5%.

10. $R = 80\% = 0.8, B = ?, P = \$76$

   $P = RB$

   $\$76 = (0.8)B$

   $0.8B = \$76$

   $\frac{0.8B}{0.8} = \frac{\$76}{0.8}$

   $B = \$95$

   The regular price of the shirt is $95.

Chapter 15 Informal Geometry

Exercise 15

1. Yes, the size and shape appear to be the same.
2. No, the size is about the same but the shape is different.
3. No, the size is different.
4. Yes, the shape is the same.
5. Yes, the shape is the same.
6. No, the shape is different.
7. 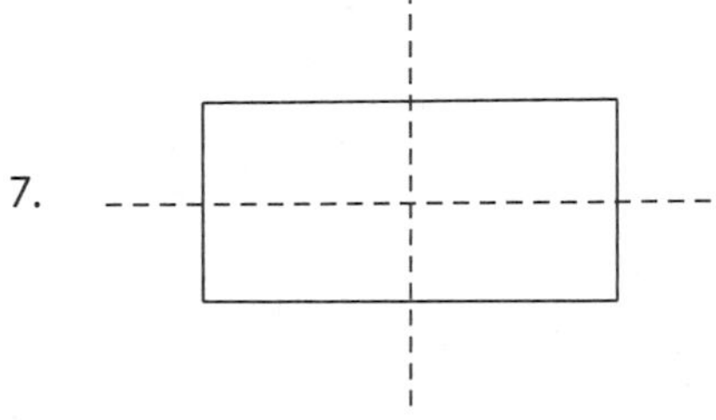

   Dashed lines are lines of symmetry.
8. The figure does not have line symmetry.
9. Acute, because 50° is between 0° and 90°
10. Right, because the angle is exactly 90°
11. Obtuse, because 130° is between 0° and 180°
12. Intersecting
13. Parallel
14. Hexagon, because it has exactly six sides

15. Heptagon, because it has exactly seven sides
16. Isosceles, because it has two congruent sides
17. Equilateral, because it has three congruent sides
18. Acute triangle, because it has three acute angles
19. Obtuse triangle, because it has one obtuse angle
20. Right triangle, because it has one right angle
21. Parallelogram, because opposite sides are congruent and parallel
22. Rectangle, because it is a parallelogram that has four right angles
23. Trapezoid, because it has exactly one pair of parallel sides
24. The length of the diameter is $2 \times 5.4 \text{ yd} = 10.8 \text{ yd}$.
25. Rectangular prism, because it has two parallel and congruent rectangular bases and rectangles for sides
26. Cylinder, because it has two parallel congruent bases, which are circles, and it has one rectangular side that wraps around
27. A square or rectangular pyramid, because it is a solid with exactly one base, triangular sides, and a square base
28. Sphere, because it is a three-dimensional solid that is shaped like a ball
29. Triangular pyramid, because it is a solid with exactly one base, triangular sides, and a triangular base
30. Pentagonal prism, because it has two parallel and congruent pentagonal bases and rectangles for sides

Chapter 16 Perimeter, Area, and Volume

Exercise 16

1. $C = \frac{1}{2}\pi(680 \text{ ft}) \approx \frac{1}{2}(3.14)(680 \text{ ft}) \approx 1{,}068 \text{ ft}$

2. The area of the floor is $20 \text{ ft} \times 30 \text{ ft} = 600 \text{ ft}^2$. Because the tiles are each of size $12 \text{ in} \times 12 \text{ in} = 1 \text{ ft} \times 1 \text{ ft} = 1 \text{ ft}^2$, 600 tiles are needed.

3. 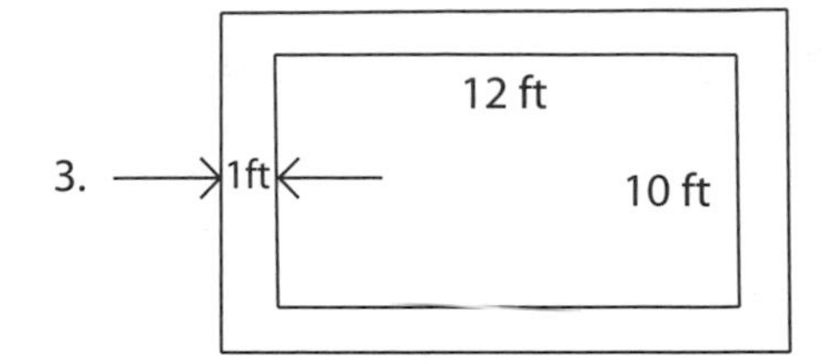

The size of the complete region is $(12\text{ ft}+1\text{ ft}+1\text{ ft})=14\text{ ft}$ by $(10\text{ ft}+1\text{ ft}+1\text{ ft})=12\text{ ft}$, so the area $A=14\text{ ft}\times 12\text{ ft}=168\text{ ft}^2$.

4. The area of the rectangle is $A_r = 4\text{ ft}\times 9\text{ ft} = 36\text{ ft}^2$. Because the square has the same area, the side of the square is $s=\sqrt{36\text{ ft}^2}=6\text{ ft}$.

5. 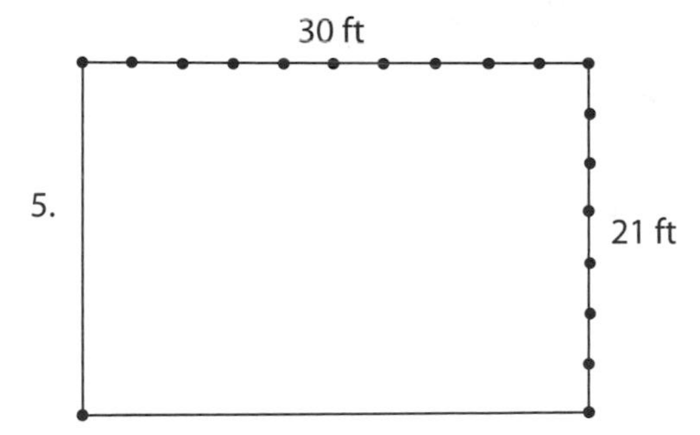

The perimeter $P = 2 \times 30\text{ ft} + 2 \times 21\text{ ft} = 102\text{ ft}$. Dividing 102 ft by 3 ft between each post gives 34. Thus, 34 posts are needed.

6. $A=\pi r^2 \approx 3.14(10\text{ ft})^2 = 3.14(100\text{ ft}^2) = 314\text{ ft}^2$

7. $V=\frac{1}{2}\left(\frac{4}{3}\pi(340\text{ ft})^3\right) \approx \frac{1}{2}\left(\frac{4}{3}(3.14)(340\text{ ft})^3\right) = 82{,}276{,}373.33\text{ ft}^3 \approx 82{,}000{,}000\text{ ft}^3$

8. a. $V=Bh=\left(\pi(20\text{ ft})^2\right)100\text{ ft} \approx \left(3.14(20\text{ ft})^2\right)100\text{ ft} = 125{,}600\text{ ft}^3$

   b. The number of bushels is $125{,}600\text{ ft}^3 \times \frac{\text{bushel}}{4\text{ ft}^3} = 125{,}600\ \cancel{\text{ft}^3} \times \frac{\text{bushel}}{4\ \cancel{\text{ft}^3}} = 31{,}400$ bushels.

   c. The value of the wheat in the full silo is

   $$31{,}400\text{ bushels} \times \frac{\$2}{\text{bushel}} = \frac{31{,}400\ \cancel{\text{bushels}}}{1} \times \frac{\$2}{\cancel{\text{bushel}}} = \$62{,}800$$

9. $A=\frac{1}{2}(21\text{ in})(14\text{ in}) = 147\text{ in}^2$

10. $V=Bh=\left(\frac{1}{2}\cdot 6\cdot 2\right)(14) = 84$ cubic units

11. Find the surface area of the box.

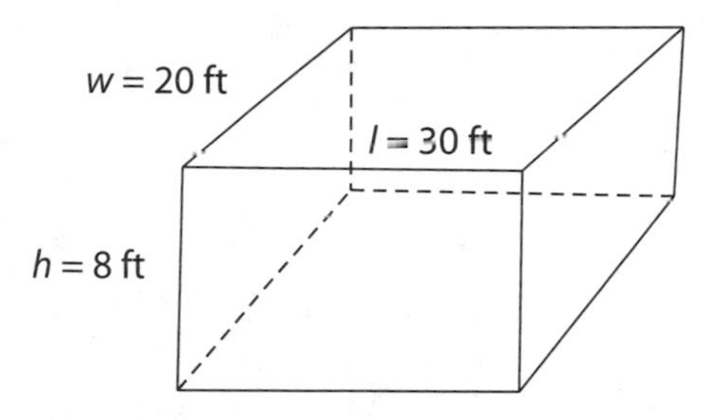

Sum the areas of the faces.

$S.A. = 2(8\text{ ft})(30\text{ ft}) + 2(20\text{ ft})(30\text{ ft}) + 2(8\text{ ft})(20\text{ ft}) = 2{,}000\text{ ft}^2$

Chapter 17 Pythagorean Theorem

Exercise 17

1. $s^2 = 27^2 + 36^2$

   $s^2 = 729 + 1{,}296$

   $s^2 = 2{,}025$

   $s = \sqrt{2{,}025} = 45$

2. $27^2 + s^2 = 36^2$

   $s^2 = 36^2 - 27^2$

   $s^2 = 1{,}296 - 729 = 567$

   $s = \sqrt{567} \approx 23.81$

3. $s^2 + 19^2 = 28^2$

   $s^2 = 28^2 - 19^2$

   $s^2 = 784 - 361 = 423$

   $s = \sqrt{423} \approx 20.57$

4. $m^2 = 25^2 + 17^2 = 625 + 289 = 914$

   $m = \sqrt{914} \approx 30.23$

5. $19^2 \stackrel{?}{=} 11^2 + 15^2$

   $361 \stackrel{?}{=} 121 + 225$

   $361 \neq 346$

   The triangle is not a right triangle.

6. The required area is the area of the semicircle minus the area of the triangle.
   Let A be the area of the semicircle and let T be the area of the triangle. Then, to find the area of the semicircle, the radius length must be known. The diameter and thus the radius can be found by using the Pythagorean theorem on the right triangle. Let d be the length of the diameter.

   $d^2 = 5^2 + 12^2 = 25 + 144 = 169$

   $d = \sqrt{169} = 13$

   The radius is $r = \frac{13}{2}$.

   $A = \frac{1}{2}\pi\left(\frac{13}{2}\right)^2 = \frac{169\pi}{8}$

   $T = \frac{1}{2}(5)(12) = 30$

   The required area is $A - T = \frac{169\pi}{8} - 30 = 21.125\pi - 30 \approx 36.37$ square units.

7. 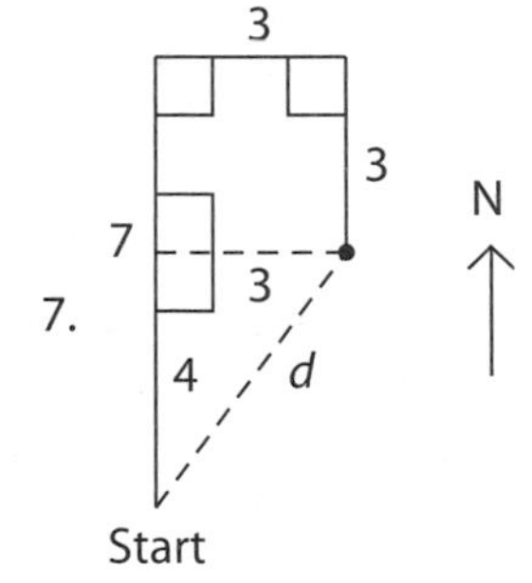

   $d^2 = 3^2 + 4^2 = 25$

   $d = 5$

   (The triangle in this figure is the special 3-4-5 triangle. Any triangle similar to this triangle has proportional side lengths, such as the 27-36-45 triangle of problem 1 earlier.)

8. The length of the sides of the square is needed for the perimeter. You can use the Pythagorean theorem to find the side length. Let s be the side length and let P be the perimeter.

   $s^2 + s^2 = 32$

   $2s^2 = 32$

   $s^2 = 16$

   $s = 4$

   $P = 4s = 4 \times 4 = 16$ units

9. Because 4 is the largest number, the question is

$4^2 \stackrel{?}{=} \left(\sqrt{6}\right)^2 + \left(\sqrt{10}\right)^2$

$16 \stackrel{?}{=} 6 + 10$

$16 = 16$

The triangle is a right triangle.

10. 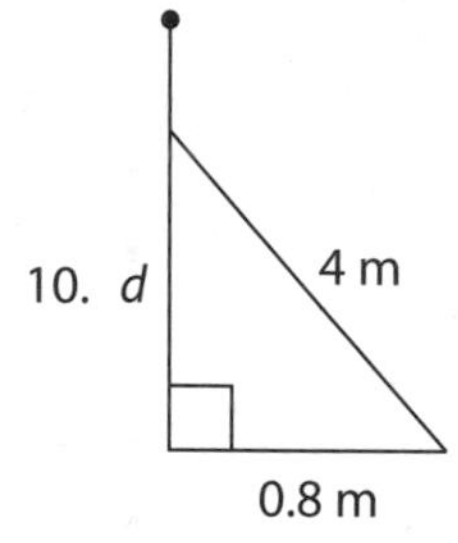

$d^2 + (0.8)^2 = 4^2$

$d^2 = 16 - (0.8)^2$

$d^2 = 16 - 0.64 = 15.36$

$d = \sqrt{15.36} \approx 3.92$

The ladder will reach up on the wall approximately 3.92 m.

Chapter 18 Counting and Probability

Exercise 18

1. no. of ways to select one door and one curtain =
   (no. of ways to select a door) $\times$ (no. of ways to select a curtain) $= 3 \times 4 = 12$ different ways
2. no. of codes possible =
   (no. of ways to select first letter) $\times$ (no. of ways to select second letter)
   $\times$ (no. of ways to select third letter) $= 15 \times 15 \times 15 = 3{,}375$ different codes
3. no. of ways to select one bread, one sandwich filling, one bag of chips, and one drink =
   (no. of ways to select a bread) $\times$ (no. of ways to select a sandwich filling)
   $\times$ (no. of ways to select a bag of chips) $\times$ (no. of ways to select a drink) =
   $3 \times 4 \times 2 \times 3 = 72$ different lunches
4. $P(5) = \dfrac{\text{number of favorable outcomes}}{\text{number of possible outcomes}} = \dfrac{1}{6}$

5. $P(\text{green}) = \dfrac{\text{number of favorable outcomes}}{\text{number of possible outcomes}} = \dfrac{14}{50} = \dfrac{7}{25}$

6. $P(\text{yellow or blue}) = \dfrac{\text{number of favorable outcomes}}{\text{number of possible outcomes}} = \dfrac{26}{50} = \dfrac{13}{25}$

7. $P(\text{2 red when drawn without replacement}) = \dfrac{10}{50} \cdot \dfrac{9}{49} = \dfrac{90}{2{,}450} = \dfrac{9}{245}$

8. $P(\text{2 red when drawn with replacement}) = \dfrac{10}{50} \cdot \dfrac{10}{50} = \dfrac{100}{2{,}500} = \dfrac{1}{25}$

9. The events are independent, so
   $P(\text{5 heads in a row}) = \dfrac{1}{2} \times \dfrac{1}{2} \times \dfrac{1}{2} \times \dfrac{1}{2} \times \dfrac{1}{2} = \dfrac{1}{32}$

10. The events are independent, so
    $P(\text{red on first spin and green on second spin}) = \dfrac{3}{10} \cdot \dfrac{1}{10} = \dfrac{3}{100}$

Chapter 19 Mean, Median, Mode, and Range

Exercise 19

1. $\text{mean} = \dfrac{15+33+30+50+0}{5} = \dfrac{128}{5} = 25.6$

2. $\text{mean} = \dfrac{-4+25-4+11+19+4}{6} = \dfrac{51}{6} = 8.5$

3. $\text{mean} = \dfrac{4.7+5.6+2.5+4.9+7.3+4.7+5.6+6.5}{8} = \dfrac{41.8}{8} = 5.225$

4. $\text{mean} = \dfrac{-10+0+3+3+6+16}{6} = \dfrac{18}{6} = 3$

5. weighted average =
   $\dfrac{10\%(80)+30\%(90)+60\%(70)}{10\%+30\%+60\%} = \dfrac{0.10(80)+0.30(90)+0.60(70)}{100\%} = \dfrac{77}{1} = 77$

6. median = 30

7. $\text{median} = \dfrac{4+11}{2} = \dfrac{15}{2} = 7.5$

8. $\text{median} = \dfrac{4.9+5.6}{2} = \dfrac{10.5}{2} = 5.25$

9. median = 3

10. median = 4
11. no mode
12. mode is –4; unimodal
13. modes are 4.7 and 5.6; bimodal
14. mode is 3; unimodal
15. modes are 4, and 10; bimodal
16. range = $50-0=50$
17. range = $25-(-4)=25+4=29$
18. range = $7.3-2.5=4.8$
19. range = $16-(-10)=16+10=26$
20. range = $500-0=500$

Chapter 20 Graphical Representation of Data

Exercise 20

1. According to the graph, 50 of the 200 women responded "No" to the survey question posed.
2. According to the graph, the median annual earnings of all full-time, year-round workers, ages 25–34, with a bachelor's or higher degree was $53,800 in the year 2015.
3. According to the graph, 29,900 (= 65% × 46,000) students enrolled at University X are freshmen or sophomores.
4. According to the graph, the small online business produced the least annual profit in 2015.
5. According to the graph, 13 of the 25 guests who attended the fundraiser for the local PAWS dog shelter were over 40 years old.
6. According to the graph, 60 of the 100 ninth-grade boys are at least 64.5 inches tall.
7. According to the graph, 10 of the 20 third graders read independently for less than 15 minutes during free time in class.

Index

Page numbers followed by *f* indicate material in figures. Page numbers followed by *t* indicate material in tables.

A

D

E

F

G

N

O

P

Z